Hunting the Faster than Light Tachyon, and Finding Three Unicorns and a Herd of Elephants

Hunting the Faster than Light Tachyon, and Finding Three Unicorns and a Herd of Elephants

Robert Ehrlich

CRC Press is an imprint of the
Taylor & Francis Group, an informa business

First Edition published 2022
by CRC Press
6000 Broken Sound Parkway NW, Suite 300, Boca Raton, FL 33487-2742

and by CRC Press
4 Park Square, Milton Park, Abingdon, Oxon, OX14 4RN

CRC Press is an imprint of Taylor & Francis Group, LLC

Library of Congress Cataloging-in-Publication Data

Names: Ehrlich, Robert, 1938- author.

Title: Hunting the faster-than-light tachyon, and finding three unicorns and a herd of elephants / Robert Ehrlich.

Identifiers: LCCN 2021056502 (print) | LCCN 2021056503 (ebook) | ISBN 9780367716257 (hardback) | ISBN 9780367708108 (paperback) | ISBN 9781003152965 (ebook)

Subjects: LCSH: Tachyons. | Neutrinos. | Light–Speed. | Relativity (Physics)

Classification: LCC QC793.5.T32 E35 2022 (print) | LCC QC793.5.T32 (ebook) | DDC 539.7/21–dc23/eng20220301

LC record available at https://lccn.loc.gov/2021056502

LC ebook record available at https://lccn.loc.gov/2021056503

A catalog record has been requested for this book.

ISBN: 978-0-367-71625-7 (hbk)
ISBN: 978-0-367-70810-8 (pbk)
ISBN: 978-1-003-15296-5 (ebk)

DOI: 10.1201/9781003152965

Typeset in Palatino
by KnowledgeWorks Global Ltd.

Contents

Preface and Acknowledgements ix
About the Author xiii

1 Three Weird Entities: Tachyons, Neutrinos, and Me 1
Hidden Unicorns 1
How *not* to Reach FTL Speed 2
$E = mc^2$: The Most Famous Equation in the World 3
"Meta" Relativity 4
You are Now Traveling at the Speed of Light 7
Phantom of the OPERA 8
My Journey through Time 10
Finding My Passion 11
Maverick Jack Steinberger 12
The Two-Neutrino Experiment 13
Good and Evil Tachyons 14
Seeing Things 15
Black Swans 17
The Neutrino as a Unicorn 18
Models of the Three Neutrino Masses 21
Other Tachyon Possibilities 23
Using Ockham's Razor Can Be Dangerous 24
Sending Messages Back in Time 24
My Message from the Future 27
Summary 28
References 29

2 Faster than Light and Backwards in Time 31
Tachyons in Fact and Fiction 31
A Lazy Dog That Could Not Find an Academic Job 32
The Block Universe and Its Worldlines 33
Challenges to the Block Universe Concept 34
Worldlines and the Light Cone 35
The Lorentz Transformation 37
Backward Time-Traveling Tachyons 38
Chasing a Tachyon 39
Searching for Tachyons 40
Messengers from Space 42
Locating the Sources of the Cosmic Rays 45

The Mysterious Cygnus X-3 ... 46
The Proton–Neutron Decay Chain ... 46
An Unsettled Mystery ... 48
A Failed Collaboration: *Glupyy Amerikanets!!* ... 48
The Field of Cosmic Ray Physics: A Work in Progress ... 50
Contacting Your Earlier Self ... 52
A "More Feasible" Way to Contact the Past ... 53
Fictional Examples ... 54
Wormhole Time Machines ... 56
Where Are the Time Travelers? ... 57
Searching for Wormholes ... 59
A Wormhole in the Solar System? ... 60
Parallels between Wormholes and Tachyons ... 62
Summary ... 63
References ... 64

3 Supernova SN 1987A and Its Three Unicorns ... 67
Supernovae ... 67
The Threat from Exploding Stars ... 68
The Birth, Life, and Death of Stars ... 69
How to Detect Neutrinos ... 72
The SN 1987A Neutrino Burst ... 73
SN 1987A and the 3 + 3 Model of Neutrino Masses ... 75
The Tachyonic Third Mass in the Model ... 77
Balancing a Seesaw with Nothing on One Side ... 78
Seeking Validation for the 3 + 3 Model ... 79
Dark Matter Holds Galaxies Together ... 80
First Confirmation of the 3 + 3 Model Masses ... 82
Revisiting the Mont Blanc Neutrino Burst ... 82
The Neutrino Detector That Failed to "Bark" ... 83
Dark Matter in the Stellar Core and 8 MeV Neutrinos ... 84
Support for the Z′-Mediated Reaction ... 86
Challenges to the Z′-Mediated Reaction Model ... 87
Finding the Unicorn Hidden in the Background ... 89
Reliability of the Background ... 91
The 8 MeV Line Revealed ... 93
Physics and the Nature of God ... 94
Summary ... 95
The Three Unicorns ... 96
References ... 97

4 Theories of Everything and Anything ... 99
Tachyons? "Sure, Why Not" ... 99
The Standard Model of Particle Physics ... 100
A Zoo of Hypothetical Particles and the Uniqueness of Tachyons ... 102

Beyond the Standard Model ... 104
What Monsters Might Be Lurking There? ... 105
Replacing Particles by Vibrating Strings and Membranes ... 105
Loop Quantum Gravity ... 107
Gerald Feinberg: An Accidental Futurist ... 109
Oliver Heaviside: An Accidental Time Traveler ... 109
Pavel Cherenkov: An Accidental Nobel Laureate ... 110
FTL Observers and Warp-Drive Spaceships ... 111
Cosmic Inflation ... 114
Dark Energy, Antigravity, and Tachyons ... 115
The Evidence for Dark Energy and an Accelerated Expansion ... 116
Mirror Universes ... 117
Other Alternatives to the Standard Cosmology ... 118
Entanglement: The Effect Einstein Found Spooky ... 119
Superdeterminism and Bell's Theorem ... 121
The Most Famous Failed Experiment in History ... 122
Einstein's Ether ... 125
Making Tachyonic Neutrinos Less Obnoxious ... 126
Logical Inconsistency of Experimental Results ... 127
Summary ... 128
References ... 130

5 Weighing the Gravitophobic Neutrinos ... 131
Introduction ... 131
Weighing the Muon Neutrino ... 132
A Scale for Measuring Imaginary Mass ... 135
The Beta Decay Spectrum and the Electron Neutrino Mass ... 137
How to Get a "Kinky" Spectrum ... 139
The Beta Spectrum Shape for a Tachyon ... 140
Neutrino Mass Experiments ... 141
An Embarrassing Episode ... 141
The KATRIN Experiment ... 143
Measuring the Electron Energy in the Spectrometer ... 145
Taking Data with Your Eyes Closed ... 146
Dealing with Controversial Results ... 147
My Overactive Imagination ... 148
A Big Letdown ... 150
Consistency of KATRIN Initial Data with 3 + 3 Model ... 150
A Stay of Execution ... 152
Hearing the Grim Reaper's Footsteps ... 152
KATRIN and Tachyons – Six Possibilities ... 153
Fitting an Elephant or a Whole Herd ... 154
Hiding Elephants ... 157
Summary ... 158
References ... 158

6 Lessons Learned **159**
A Bright Spot in the Darkness 159
A Third Approach to Physics Research 162
Pros and Cons of Data Prospecting 165
Negative Evidence and Reviews 167
Fake or Predatory Journals 168
Making as Many Mistakes as You Can 169
Spotting Promising Anomalies 172
Search for Extraterrestrial Intelligence (SETI) 173
Dyson Spheres 177
Are We Alone? 178
Our Chances of Making It to a "Post-Human" Era 179
Are You Living in a Computer Simulation? 181
Guessing the Odds 183
How Could We Tell? 185
My Belief That I Am Living in a Simulation 186
Some Concluding Thoughts 188
Summary 189
References 190

Index **191**

Preface and Acknowledgements

This book is a scientific detective story. The crime is *speeding*, that is the breaking of the cosmic speed limit of 299,792,458 meters per second, which is the exact value of the speed of light in vacuum. Any perpetrator violating that cosmic speed limit would be committing a very serious crime according to nearly all physicists – perhaps the worst crime in over a century, if not ever. Unlike most detective stories, where the mystery is the identity of the perpetrator, here the main uncertainty is whether the crime has in fact been committed by a well-known subatomic particle, the neutrino. Of course, physicists continually discover new particles, so it is also possible that some as-yet-unobserved species of particle is the guilty party. Whether the speed violator is a neutrino or some new particle it would be called a *tachyon.*

This detective story is also a memoir written by a "tachyon hunter," who has been pursuing these beasts for over two decades. My pursuit of them has been in the face of extreme skepticism from most of my physics colleagues, who point out that extraordinary claims require extraordinary evidence. The skeptics also note correctly that no superluminal neutrinos or other particles have ever been observed, and there have been many false sightings. Despite that history, considerable circumstantial evidence has been published in physics journals that some neutrinos are probably tachyons. The term *circumstantial* is used here exactly the way it is used in law, meaning evidence that relies on an inference to connect it to a conclusion of fact. Moreover, it has long been recognized that some circumstantial criminal evidence, such as DNA on the murder weapon, can be considerably stronger than direct evidence, such as eye-witness testimony of the murder.

As you may know, Albert Einstein, in his 1905 special theory of relativity, prohibited faster than light particle speeds. Fifty-seven years later three physicists, O. M. P. Bilaniuk, V. K. Deshpande, and E. C. G. Sudarshan rediscovered a loophole in relativity. The trio noted that Einstein forbade superluminal speeds only for particles that were initially moving *slower* than light. Therefore, hypothetical tachyons would not conflict with relativity provided that, from the moment of their creation, they *always* moved *faster* than light. In this case, the speed of light, instead of being an upper limit to speed, would represent a two-way barrier. Normal matter like electrons, protons, or human beings, would forever be on the slower than light side of the barrier, and tachyons would forever be on the faster than light side.

Even though we slower than light creatures could never catch up to and capture a tachyon, that does *not* mean we could not observe its superluminality. If a single subatomic particle were found to have traveled a known distance from its source to a detector in less time than light, the particle would be a tachyon. As we have noted, there already is a good tachyon candidate among one of the well-known particles, namely the "ghostly" neutrino. Neutrinos are special because their measured speed has always been found to be very close to that of light. Second, there are reasons to believe their speed slightly differs from light, making them either a bit faster or slower. Even if only a handful of neutrinos were found to move a tiny bit faster than light, it would be a very big deal in the world of physics. It would be a much bigger deal than when airplanes first broke the sound barrier, which some "experts" thought was impossible.

Should you find yourself in the "Who cares?" category, I would note that if neutrinos are tachyons, the assumptions of theoretical physics would be turned upside down. Even the order of cause and effect could be switched. Suppose, for example, Alice sends Bob a message using a beam of tachyons. She could do this by simply modulating the beam into a series of long and short pulses using Morse code. Relativity tells us that some observers would say the tachyon message really went from Bob to Alice, and that we cannot say who is right. Furthermore, such a switch between sender and receiver of a message might even allow Alice to send messages to her earlier self. For now, these wild speculations remain in the category of science fiction unless the existence of tachyons should be surely demonstrated in an experiment.

An ongoing experiment in Germany known as KATRIN might conceivably provide such evidence. This experiment is designed to measure the mass of neutrinos rather than their speed. If neutrinos really are tachyons, then their mass would be an imaginary number. Note that "imaginary" is here used in its mathematical sense, meaning that the square of the neutrino mass is negative. Imaginary mass particles would have the extremely weird property of speeding up when they lose energy. In fact, on losing all its energy, a tachyon would be moving at infinite speed, and travel through the entire universe in no time at all. While the KATRIN experiment's initial results have not shown neutrinos are tachyons, nor do they exclude the possibility. Eventually, KATRIN may yield a definitive result on the matter, however, as the experiment accumulates more data, and the measurement uncertainty in the neutrino mass shrinks. Even if the experiment ultimately fails to show that one (of three) types of neutrinos are tachyons, it will not thereby prove that tachyons do *not* exist. All that a negative result can do is set limits of some kind, whether the search is for tachyons, extraterrestrial civilizations, or hidden extra dimensions.

Tachyons are just one of many crazy ideas in science, some of which defy expectation and turn out to be true. My affinity for *crazy ideas in science* has motivated much of my research and led me to write several books on that very theme for the general reader. All that the present book will require of you, aside from skills and knowledge you probably learned in high school, is a deep curiosity about our very strange universe and an openness to unconventional ideas. In many places, the book crosses the blurry line between science fact and science fiction. Doing so is appropriate, because ideas gleaned from science fiction have often led physicists to research subjects such as tachyons, extraterrestrials, and time travel.

I wrote this book as a retired 83-year-old physics professor who is nearing the end of his journey through time. In my remaining time on this planet before the atoms of my body return to the stardust from whence they came, I can only hope that I live to see a definitive result from the KATRIN experiment. Even if that experiment yields the result nearly all physicists expect, namely no evidence of neutrinos being tachyons, I believe that my scientific detective story should be of interest to many readers. As the story unfolds, we shall come across important lessons for anyone interested in pursuing their own unconventional ideas in science. To accompany the book, I have created a fun and a bit tacky (tachy?) website: *Ehrlich.physics.gmu.edu*. Whether you are a novice or an expert you will find much to explore there, including a list of over eighty questions (with answers) about tachyons and time travel, and many links to various resources on those two closely related subjects.

I am indebted to all the people who reviewed sections of this book and made helpful suggestions, including John Todd and Dublin artist Sinead McDonald. Many of my George Mason University colleagues were also helpful in reviewing the book including Political Scientist Jim Pfiffner, Space Scientist Art Poland, Geologist Doug Mose, Astronomer Harold Geller, and Science Communication specialist Kathy Rowan. Their comments helped me understand which sections needed extensive work to make them more understandable to non-scientists. Several reviewers including former Mason colleagues, Economist Jim Bennett, Historian Peter Stearns and retired Librarian Ian Fairclough made detailed stylistic and substantive comments on the entire book. Jim even reviewed two drafts of the book. I especially thank my friend of over 60 years physicist Donald Gelman for his tremendously helpful comments.

Many other physicists were extremely helpful during my research on tachyons. They include Mason physics colleagues Maria Dworzecka and Robert Ellsworth, who were appropriately skeptical of some of my half-baked ideas. My collaborations with Man Ho Chan and Ulrich Jentschura led to the publication of three (out of a total of 18) articles about tachyons, which I could not have written alone. Special thanks go to Alan Chodos: his help in better understanding the physics of tachyons has been invaluable. Alan's work with Alan Kostelecky, Robertus Potting and Evalyn Gates on neutrinos as tachyons got me initially interested in the subject, and it was in part his encouragement that led me to write this book. Alan has also very kindly agreed to handle the final editing in the event of my death or incapacitation.

Although my tachyon research has not involved me in any experiments, most of my work has relied on analyses of experimental work published by others. The work of three specific groups has been most helpful. First, I am indebted to Masayuki Nakahata and the other members of the Kamiokande Collaboration who detected the neutrino burst from supernova SN 1987A. The inclusion in their 1988 Physical Review article of neutrino data taken in the hours before and after the main burst, was invaluable in discovering important evidence showing that some neutrinos may be tachyons.

Second, I am grateful to Antonio Eridato and the other members of the OPERA group at CERN, who announced their "neutrino anomaly" to the world in 2011. They did so even though they knew from the outset that it probably was not truly due to faster than light neutrinos. The neutrino anomaly, even though it later proved to be a phantom of the OPERA, reawakened in me an interest in the subject that I had studied over a decade earlier.

Finally, I especially wish to thank the KATRIN Collaboration, which has been conducting an incredible experiment to measure the electron neutrino mass with far higher precision than previous measurements. Group leader of KATRIN, Guido Drexlin, has been particularly kind to me. Diana Parno, one of the people in charge of their data analysis, and Hamish Robertson were also very kind in keeping me informed of the experiment's status. I particularly appreciate Diana's willingness to discuss details of the data analysis with me, as well as her reassurance that they were taking my prediction for the experiment seriously. I am also very thankful to Hamish Robertson for reading and correcting several errors in Chapter 5 dealing with the KATRIN experiment. This experiment could not have come along at a better time, and I thank the KATRIN people for inviting me to give several seminars at KIT. I am also grateful that they did not treat me as a crackpot: even though I suspect that they were extremely dubious about my prediction for the experiment's results.

I want to thank all the reviewers of my tachyon papers, especially those who criticized them severely and often recommended against publication – thereby giving me the time and incentive to improve them. In a similar vein, I am indebted to someone or something responsible for a mysterious seemingly automated message I received on November 22, 2015. I seriously wondered if this could have been a message from the future. Whatever its source, I interpreted the "message" as a sign that I was on the right track in my research, and I should keep at it. Most of all, I thank my wife Elaine for being my best friend and soulmate for 60 years and counting, and for her serving as a good sounding board for my ideas.

Finally, I thank God, about whose existence I am devoutly agnostic, for putting all those clues out there suggesting that neutrinos might be tachyons. Through those clues I believe that I was able to see through the excellent "camouflage" that allowed tachyons to remain hidden for over a century. Alternatively, if the KATRIN experiment should not support my prediction for it, and it has all been a magnificent delusion, I can only admire God for his great sense of humor and interpret His actions as being merely mischievous, not malicious. As Einstein once said, "Subtle is the Lord, but malicious He is not."

About the Author

Robert Ehrlich is an American physicist who holds a PhD in physics from Columbia University, where he participated in the Nobel prize-winning muon neutrino experiment. He retired from George Mason University in Fairfax, Virginia, having served 15 years of his 36 years there as physics and astronomy department chair. Dr. Ehrlich is a fellow of the American Physical Society and his primary area of scholarship is particle physics. He is also well known for his contributions to science education, particularly with simple physics demonstrations, renewable energy, and computer applications in physics education. His 20 books have been translated into six languages.

1

Three Weird Entities: Tachyons, Neutrinos, and Me

Physics advances by accepting absurdities. Its history is one of unbelievable ideas proving to be true.[1]

Rivka Galchen, Novelist

Hidden Unicorns

Suppose you received a message appearing to be from the future – maybe even from someone claiming to be your future self – what would it take for you to believe it was not a hoax? This question may not be just science fiction if Faster Than Light (FTL) particles known as tachyons exist, since they might allow messages to be sent back in time. It is true, of course, that earlier reports of FTL particles have proven mistaken, and many physicists think they are impossible. Nevertheless, considerable published evidence suggests that tachyons may exist.

Tachyons might be thought of as the "unicorns" of physics – either too bizarre in their properties to be anything but mythical creatures, or if real, then too well hidden for any convincing evidence of their existence to have emerged. I argue in this book that they have, in fact, been well hidden because of some exceptionally good camouflage. The reader will discover how I and other physicists believe we have seen through the camouflage, and gradually uncovered evidence for the existence of tachyons. More precisely, we have published evidence that some neutrinos, a well-established species of subatomic particles, are tachyons. If *confirmed*, such a finding would be entirely consistent with the ghostly neutrinos revealing their properties only gradually and having them come as one surprise after another.

Most physicists, however, remain extremely skeptical because tachyons would violate the principle of Einstein's Theory of Relativity which prohibits superluminal (or FTL) speed. Even worse, tachyons erase any absolute distinction between cause and effect, which in physicist-speak is a violation of

DOI: 10.1201/9781003152965-1

causality. *Most importantly, no completely decisive empirical evidence for neutrinos being tachyons has yet been found.* However, much evidence for their existence is out there if one lets the data speak for themselves without imposing preconceived assumptions. You can think of this book as the legal brief presented by a prosecutor who is convinced that some accused neutrinos are guilty of violating the cosmic speed limit, while he is also aware that it is not an open-and-shut case, and there is much reluctance on the jury's part to convict. Fortunately, an ongoing experiment known as KATRIN could yield the decisive confirmation – or alternatively it could show that tachyons remain suspended between science fact and fiction. Writing about a potential discovery before it is made may seem idiotic but as an 83-year old with a limited time horizon I thought it might be wise to write this book while I still could. Part of me even imagined that by putting these words on paper I could somehow bring about the desired outcome of the experiment, which I know is logically impossible.

In addition to describing evidence for some neutrinos being tachyons, this book describes an unfamiliar method for making scientific discoveries. Many important discoveries arise from accidental observations, as was the case with penicillin, superconductivity, and radioactivity. More commonly, they are the result of a theory such as relativity or the wave theory of light, which then is confirmed by a single dramatic observation. Here a third method called *data prospecting* is described. Data prospectors look for interesting patterns or anomalies in previously published data, and then try to create an unconventional model to explain them. Finally, they seek confirmation for the model in a wide variety of other published data sets, while always seeking negative evidence that could doom their model.

The book has been written for the general reader who is curious about the physical universe, and in awe of its magnificence. If you have taken a physics course or read popular books on the subject, that would be very helpful, but it is not essential. Even without a physics background, you should still find most of the book understandable provided you are willing to wrestle with some challenging concepts. On the other hand, if you are expecting a "woo-woo" story outside the bounds of physics, you will likely be disappointed.

How *not* to Reach FTL Speed

Einstein's prohibition of FTL speed particles does seem to violate common sense. For example, imagine that you are in an interstellar spaceship far from Earth, and the ship engines had a constant thrust that accelerated it at "one g." With that acceleration if you started at rest then after each passing second people back home would judge your speed to increase by about ten meters per second. One benefit of the one g acceleration is that as the ship accelerated you would feel like you are back on Earth with your normal weight instead of floating around.

Here is some very simple math. With a constant rate of increase in speed of ten meters per second, the ship would attain a speed of 10, 20, and 30 meters per second after 1, 2, and 3 seconds. You might imagine it would reach the speed of light, which is about 300 million meters per second in 30 million seconds, or just under a year. Einstein, however, proved that this simple calculation is in error. That's because as the ship's speed increases its constant engine thrust would cause less and less gain in speed each second, because Einstein showed that the ship's mass would increase with its speed as judged by Earth observers. In fact, as the ship's speed approaches that of light the mass would increase without limit, thereby making light speed unattainable. On the other hand, you a passenger aboard the ship, would not notice the mass increase of the ship or yourself, because you could consider the ship to be at rest.

$E = mc^2$: The Most Famous Equation in the World

The increase in mass with an object's speed reflects the fact that mass is a form of energy – that being one meaning of Einstein's famous equation $E = mc^2$, where c is the speed of light. In addition, $E = mc^2$ tells us that even objects at rest have an energy inherent in them, which is in proportion to their mass. You can think of the c^2 in the equation as the "exchange rate" telling us how much energy is equivalent to one unit of mass. Given the huge size of c^2, a tiny amount of mass is equivalent to a huge amount of energy. The equation $E = mc^2$ implies that in any reaction where energy is liberated (such as combustion), there will necessarily be a decrease in the total mass. Einstein's equation was made famous because of the atomic bomb, where a huge amount of energy is released, but it also applies to ordinary chemical processes where the change in mass is too small to notice, given the huge value of c^2.

Einstein was a 26-year-old patent clerk in 1905, when he devised his theory of special relativity. The equation $E = mc^2$ emerged as an "afterthought" put forward in a second relativity paper published a few months after the first one. A photo of him taken from this time was used for a postage stamp shown in Figure 1.1, one of at least 149 Einstein stamps worldwide – more than any other scientist. In case you're curious, Marie Skłodowska Curie, the only person to have received Nobel Prizes in two different scientific fields (chemistry and physics) is in second place, having appeared on 65 stamps.

FIGURE 1.1
A young Albert Einstein circa 1905 on a postage stamp.

Source: Image from Shutterstock

"Meta" Relativity

Despite Einstein's prohibition of FTL speeds, in 1962 three physicists figured out a loophole that might allow them to exist. This loophole was discussed by O.M.P. Bilaniuk, V.K. Deshpande, and E.C.G. Sudarshan in their paper titled "Meta" Relativity.[2] The primary author of this paper, Indian-American physicist George Sudarshan, was one of the giants of 20th century physics – see Figure 1.2. Although he never was awarded the Nobel Prize, Sudarshan was nominated for it nine times, which apparently is not especially unusual. The record-holder for the most nominations for the physics Nobel has been Arnold Sommerfeld who received 84 nominations during his lifetime without ever receiving the prize, and Einstein himself was not awarded his Nobel until his eleventh nomination. Incidentally, Sommerfeld is the first person who considered FTL particles a year before Einstein's relativity paper, and he named them "meta-particles."

Sudarshan and his coauthors of the 1962 "Meta" Relativity paper realized that Einstein's ban on FTL speed applied only to particles such as electrons that were accelerated continuously over some time interval up to the speed of light, c, which Einstein proved in 1905 would require an infinite amount of energy. In fact, Einstein had specifically limited his ban on FTL speeds to the case where particles initially had a sub-light speed and were gradually accelerated. He did not consider the possibility of particles being instantaneously created in subatomic collisions. When this occurs, the created particles might

FIGURE 1.2
Indian physicist E. C. G. (George) Sudarshan in 2009.

Source: Photo by Tabish Qureshi

have FTL speeds at the very instant of their creation, without any infinite energy being required to bring them up to light speed.

The mathematical trick used by Sudarshan and his colleagues in their paper was to require hypothetical FTL particles to have a mass that had an imaginary value, meaning that the mass *squared* (m^2) is negative. What could such a bizarre idea mean? Using Einstein's equation for the variation of a particle's mass with speed, negative mass *squared* means the particle would have a real measurable energy only if its speed is always FTL or *superluminal*. In addition, an imaginary mass particle would speed up as it loses energy, and its speed would approach infinity when its energy approached zero, as depicted in Figure 1.3. Despite such strange properties, the "Meta" relativity authors noted that "imaginary mass particles offend only the traditional way of thinking." However, for those who are easily offended, there is an alternative version of their theory which avoids imaginary masses entirely.

Throughout the remainder of this book the word *mass* refers to the so-called proper or ***rest*** mass of a particle, even though FTL tachyons could never be at rest. Tachyons, however, are no weirder in this regard than photons (or particles of light), which also can never be at rest, and always have a speed c in vacuum. When we previously took an imagined spaceship ride, it was not the proper mass that increased with speed but what is usually referred to as the *relativistic mass* – a quantity that is proportional to an object's energy.

If we assume that mass can be imaginary, c becomes no longer an upper limit to speed but rather a two-way speed limit, so that particles having higher speeds can never descend below it. This two-way limit neatly divides

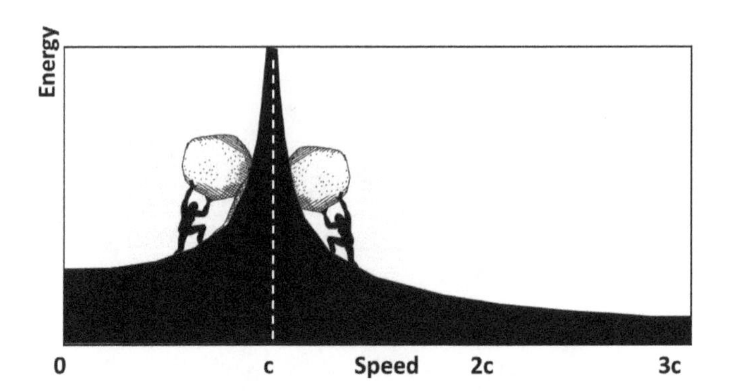

FIGURE 1.3
The height of the hill shows the energy of a particle versus its speed together with the mythical Sisyphus trying to push the boulder up a hill getting progressively steeper, making c a two-way speed limit

all the matter of the universe into three classes known as bradyons, tachyons, and luxons. *Bradyons* (sometimes called tardyons) are ordinary particles like electrons that always have sub-light speed and positive m^2; *tachyons* always have FTL speed and negative m^2, and *luxons* always have a speed exactly equal to light and zero m^2. The photon may not be the only luxon, assuming the hypothetical graviton exists, as a particle that transmits the force of gravity. In this tripartite classification scheme, infinite energy would be required for a particle in any one of the three classes to move into another, which therefore could not happen. Thus, for example, since tachyons gain energy as they slow down, they would need to gain an infinite amount of energy to slow down to the speed of light.

Obviously, since hypothetical tachyons can never have a speed less than light, we could never catch up to one. You might, therefore, imagine they could never be observed but that is not the case, as we shall see. Remember, however, that the phrase "speed of light" refers here only to the speed of light in vacuum, c, and not that in a transparent medium such as glass, air, or water, in which light travels at a reduced speed. As a result, bradyons like electrons could exceed the speed of light in such media, and they have been routinely observed to do so using particle detectors, which incidentally can (with an appropriate app) include your smart phone.

Ever since tachyons were first proposed in 1962 researchers have searched for them in experiments. Three ways to look for tachyons would be to look for particles whose measured mass is imaginary (or mass-squared is negative), whose speed exceeds c, or whose speed increases as they lose energy. There are, however, at least five additional ways to search for tachyons. While some of these searches initially yielded positive results, they either were one-time observations or flawed experiments. In fact, the Particle Data Group, which annually compiles summaries of results of all subatomic particle properties, got so tired of seeing false reports of tachyons that since 1994 they stopped including them.

You are Now Traveling at the Speed of Light

As we have seen, the idea that an infinite energy is needed to bring a particle up to light speed can explain why the speed c represents an upper limit for ordinary matter, that is, bradyons. However, some physicists have an alternative explanation of why c is a universal speed limit. This explanation makes the startling claim that we are *always* traveling at the speed of light! This strange-sounding idea depends on the proposition that motion can refer not only to motion through space but also through time. If we are at rest in space, our speed through space is of course zero, and our speed through time is at its usual rate of one second per second. However, if we move through space, someone watching us go past would, according to relativity, judge our motion through time to be reduced (slowed down) in such a way that our net speed through space and time remains unchanged. These ideas are clarified in Figure 1.4, which shows how the relationship between the motion of an object through space and time can be related as points lying on a circle.

For example consider eight specific cases: A, B, C, … H. Object A is stationary and moves only through time at the usual rate, while object B moves equally through space and time. According to relativity, such an object's motion through space would be 70.7% of the speed of light, and the rate of passage through time would be 70.7% of the usual rate, in other words for such an object time slows down (or "dilates") by this factor. The object's combined or net speed through space and time (found from the Pythagorean Theorem for the yellow triangle) works out to be just c, the speed of light.

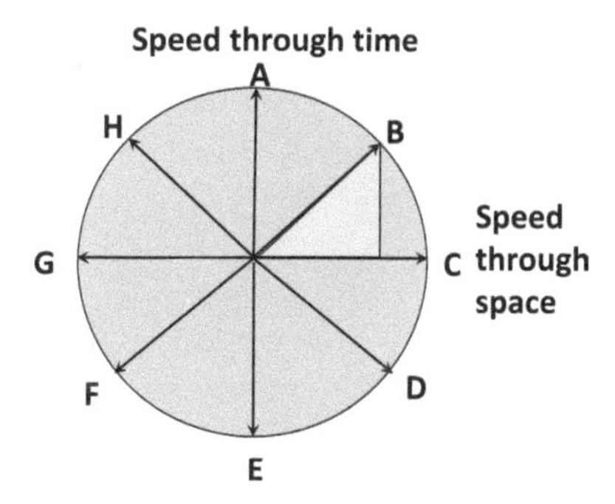

FIGURE 1.4
Relationship between any object's speeds through space and time, with eight examples. For case B, the speeds through space and time are equal.

Object C would correspond to a photon which moves only through space at speed c, and for which time essentially stands still. If you were able to move with a photon as it traveled from a distant star to Earth, you would complete the journey in no time at all. Objects represented by F, G, and H would be moving to the left through space instead of to the right, while objects D, E, and F would be traveling backwards in time, if such a thing were possible. You may find this explanation of why the speed of light is an upper limit to the speed of normal matter less persuasive than the previous infinite energy argument, but both explanations can be shown to be equivalent, with one important difference. The notion that particles always move at the speed of light would leave no room for FTL tachyons.

Phantom of the OPERA

Among all the known particles, only neutrinos have a chance of being FTL tachyons. The privileged role of neutrinos stems from their tiny masses. Their measured masses are so close to zero that it is uncertain whether their mass *squared* is positive or negative. Another related reason neutrinos are special is that when their speed is measured instead of their mass, their speed v is always so close to c that it is uncertain whether the difference, v – c, is positive or negative.

Given the lack of conclusive evidence for the existence of tachyons, or for neutrinos being tachyons, one might think that interest in the subject would gradually wane, but as Figure 1.5 shows, the opposite is true. There has been a roughly exponentially rising number of physics papers with "superluminal"

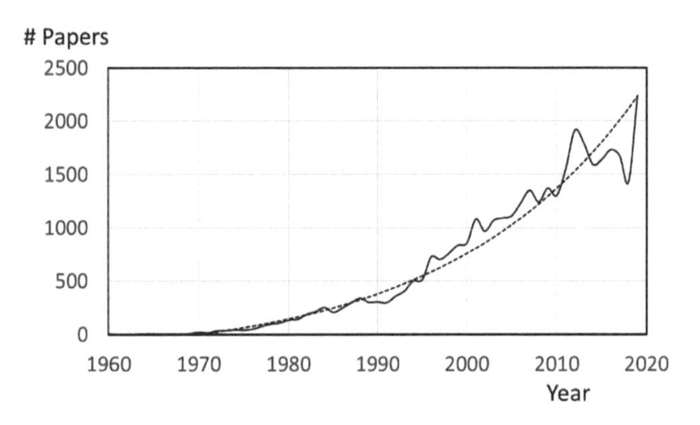

FIGURE 1.5
Number of papers by year containing "superluminal" in the title or abstract and a smooth exponential curve.

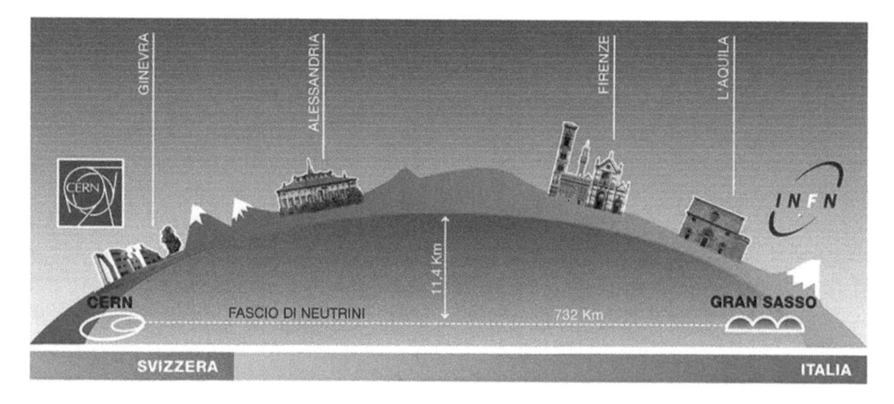

FIGURE 1.6
An experiment tracked how long neutrinos took when they traveled from CERN through the Earth to reach a detector 730 km away in Gran Sasso, Italy.

Source: Credits ©INFN

in their title or abstract since tachyons were first proposed in 1962. As already noted, there have also been many false claims of FTL tachyons, most recently in 2011. At that time the OPERA group at the CERN accelerator in Geneva reported that a beam of neutrinos was measured to have a speed very slightly FTL – see Figure 1.6. Although the excess above light speed was only 0.0000237%, it was well outside the uncertainties of the experiment. Given the potentially Earth-shattering consequences of the claim, the original measurement was repeated, after several problems were identified including a loose cable. In the repeated measurement, the corrected neutrino speed was indistinguishable from that of light. The difference between the measured and expected arrival time of neutrinos (compared to that of light) was approximately 6.5 ± 15 billionths of a second, which of course, neither requires nor excludes a very slight superluminality. Putting it another way, the result did not show the neutrino speed was surely below that of light.

The 2011 OPERA retracted claim of superluminal neutrinos may have had an impact on the number of papers on the subject, as Figure 1.5 shows. In 2012 there was a rapid rise in their number, followed by a temporary fall for a few years, after the initial claim of FTL neutrinos proved a "phantom." Despite many such mistaken claims of FTL particles over the years, some physicists (including me) have never given up hope that definitive evidence for tachyons might be found. However, even for aficionados, it is misleading to say we believe in tachyons, because science is all about evidence, not belief, and this book describes that evidence. Some of the evidence discussed relates to a property of tachyons that has made them a source of special fascination. I refer here to the connection between FTL speed and traveling backward in time, as exemplified by a well-known limerick by A. H. Reginald Butler about a young lady named Bright.[3]

My Journey through Time

All of us are time travelers in our own minds as we access memories or imagine the future. We are also actual time travelers as we "row, row, row our boat" down the stream of time. As an octogenarian, I am now near the end of my journey through time, but I remember well that young lad I once was, who mentally still lives within. That boy developed an interest in science primarily through reading popular science books, and especially by watching an amazing TV program featuring Don Herbert aka "Mr. Wizard" who involved both boys *and* girls in simple experiments, some of which produced astonishing results. Best of all, most of these experiments could be recreated by viewers. The Mr. Wizard shows, which can still be found on YouTube, originally ran from 1951 to 1990, and it has been said by Bill Nye, another science popularizer, that no other fictional hero has rivaled Mr. Wizard's popularity and longevity. Don Herbert has been credited through his inspiration of helping to create a first generation of rocket scientists in the United States – a group that was responsible for the successful quest to reach the moon. I have always tried to emulate Don Herbert in my own teaching by using many simple demonstrations to illustrate difficult concepts. Now that I am retired and no longer teaching, I still enjoy speaking to groups (especially kids), and occasionally putting on a "science magic show," when I can sneak in some physics as well as entertain – see Figure 1.7.

Among the sciences, I was initially drawn to chemistry, given my fascination with explosives. A special note to the authorities: this was done without any terroristic intent, and no humans or animals were harmed during these "investigations" usually carried out in a nearby vacant lot. The closest I came to disaster was when I obtained some metallic sodium, which reacts violently with water. While I was fooling with it in the bathroom of our house, I got a small piece wet which set the shower curtains on fire. My mother came in with a bucket of water to put out the fire, which unfortunately caused a much larger fire when the water reacted with more of the sodium. Somehow, we got the fire under control before the water reached the main chunk of sodium, which might well have had very unfortunate results. My mother then marched me and the sodium out to a vacant lot to bury it in a hole mom made me dig. Of course, the ground was damp, and following the resulting explosion an interesting encounter occurred with a fire marshal who amazingly just happened to be passing by.

Remarkably, in those days, before the concerns of terrorism and liability, kids were able to obtain all sorts of dangerous chemicals – either in chemistry sets or directly from chemical companies, as I had done with the sodium. It was even possible to buy radioactive ores, such as in the "Gilbert U-238 Atomic Energy Laboratory." Many decades later, the Gilbert lab set was criticized as "the world's most dangerous toy" because of its radioactive material. It was pulled from the market after only two years but the manufacturer

FIGURE 1.7
The author performing a science magic show for kids (of all ages) in his community. The globe is a "plasma globe."

never acknowledged that the radioactive material in each set might be harmful. Gilbert did, however, apparently note that users should not break the seals on the ore sample jars, because they would run the risk of having radioactive ore spread out in "your laboratory." Whether the risk was that your mother would be upset or that your "lab" would become dangerously contaminated, and you would die was not specified, however.

I never owned an Atomic Energy lab set, but one of my first jobs while still in college was at the former Atomic Energy Commission (AEC), now the Department of Energy. I remember holding the warm-to-the-touch plutonium core of a nuclear bomb, whose intense radioactivity could not penetrate the outer layer of my skin, being of the alpha variety. I also recall being greatly disappointed when my visit to a bomb test was cancelled due to the signing of the nuclear test ban treaty – so much for my priorities as a young adult.

Finding My Passion

After my initial flirtation with chemistry, I was drawn to the kind of science that explains how things work at some fundamental level, which pretty much meant physics. Ernest Rutherford, who discovered the atomic nucleus in 1911, has been quoted as claiming that "all science is either physics or stamp collecting." By this disparaging remark, he meant that most fields of science like geology, entomology, and botany are essentially concerned with giving names to things and classifying them into groups, not unlike what stamp collectors do. Of course, Rutherford whose main work was during the early years of the 20th century, came before revolutionary developments in many of the sciences (especially biology) that transformed some of those disciplines into anything but mere classification schemes. One may also question Rutherford's characterization of other sciences when his assessment of the future of his own science was so off the mark – just consider this remark he made before the British association for the advancement of science: "Anyone who expects a source of [nuclear] power from the transformation of these atoms is talking moonshine."[4] Rutherford has also been credited with noting that "If you can't explain your physics to a barmaid, it is probably not very good physics."[5] Apart from the sexism of this remark, I find myself more in sync with it than his barb about other sciences being merely stamp collecting. While the mathematics of "good physics" may be abstruse, its essence is usually inherently simple.

Apart from my boyhood fascination with explosives, I also enjoyed taking things apart *non*-explosively to see how they work. However, my efforts to reassemble the watch or other device I had taken apart usually were stymied, and I wisely concluded that I had better avoid the more practical field of engineering (my first major in college) in favor of physics. Anyway, physics was a much better fit for me given my interest in understanding the "why" behind things, a matter of little concern to most engineers.

Physics is divided into subfields such as optics, atomic physics, and nuclear physics, with most physicists working entirely within one specialty. In my case that was particle physics, with my first foray into this field being the famous "two-neutrino experiment," when I was a PhD student at Columbia University. Prior to this experiment, neutrinos created in all processes were thought to be the same particle. However, this experiment showed that neutrinos associated with electrons were different from those associated with particles known as muons (sort of ultra-heavy electrons). The two-neutrino experiment resulted in Columbia University professors Leon Lederman, Melvin Schwartz, and Jack Steinberger sharing the Nobel Prize for their discovery. My involvement in the experiment was at the behest of my PhD thesis adviser Jack Steinberger.

Maverick Jack Steinberger

Jack had been at Columbia since 1950 after leaving the University of California at Berkeley, where despite his achievements, he was told to leave for his refusal to sign a Non-Communist Oath. Jack, never a Communist, always disliked conformity and being told what to do, especially in the political or religious realms. Raised in a Jewish family in Germany, Jack had emigrated to the United States in 1933 at age 13, the year Hitler came to power. Although identifying as Jewish, he once confessed to an interviewer "I'm now a bit anti-Jewish since my last visit to the synagogue, but my atheism does not necessarily reject religion."[6] After getting a bachelor's degree in chemistry and serving in the United States Army Signal Corps, he did his graduate work under the great Enrico Fermi. Like Steinberger, Fermi also migrated to the United States after fleeing fascism in Europe, but in his case, the year was 1938, the same year he won the Nobel Prize, and the fascist dictator of his native Italy was Mussolini not Hitler. Steinberger, after leaving Columbia, pursued his research at CERN, where he remained active in physics until his passing at age 99.

LOYALTY OATH

In 1950, the state of California passed a law requiring all state employees to sign a loyalty oath that disavowed radical beliefs, specifically membership in the Communist Party or "any party or organization that believes in, advocates, or teaches the overthrow of the United States Government, by force or by any illegal or unconstitutional means." Several professors at the University of California resigned in protest or lost their positions when they refused to sign the oath. Those who left in protest included three German Jewish refugees from Nazi Germany. In total, the University Regents fired 31 faculty members who refused to sign the oath. One of the fired faculty members, physicist David Saxon, was later appointed President of the entire University of California system in 1975.

The Two-Neutrino Experiment

The two-neutrino experiment, also called the muon neutrino experiment, was my first research experience as a graduate student. In contrast, many of today's physics students engage in research while undergraduates. At the time, I had little appreciation for the importance of this experiment, and my role in it was relatively minor, as compared to that of the other three graduate students involved. Although my "grunt" work was insufficient to merit co-authorship of the paper reporting its results, I was acknowledged in the paper for my "computation of neutrino cross sections." Looking back now some 60 years later, I still fondly remember the many nightshifts I took during experimental runs, in which I essentially served the role of a glorified baby-sitter. I recall one night when Steinberger stopped by to inquire what I was doing, and I responded that I was watching the gauges on an instrument panel. However, when confronted with his natural follow-up question of why I was doing that, I responded inanely that I would call someone if the readings changed significantly indicating a possible problem.

Despite the smallness of my role in the two-neutrino experiment, my involvement planted the seed that perhaps I too might someday follow my adviser Steinberger, and his adviser Enrico Fermi, to make an important discovery. Physicists are usually divided into either experimentalists or theorists, with very few able to make significant contributions in both areas – Fermi being one of the few notable exceptions. It is interesting that Fermi essentially migrated from theory to experiment, because his seminal theoretical paper on the proposed weak interaction was rejected by the prestigious journal *Nature* because "It contained speculations too remote from reality to be of interest to the reader."[7] Fermi is also noteworthy because he was one of eight scientists whose rejected papers laid the basis for their Nobel Prize. Fermi's experience makes it easier for the rest of us to imagine that our rejected papers really were brilliant, and they were rejected merely because of the idiocy of a reviewer or editor. Alas, I must confess that in retrospect the rejections of my own papers were justified much more often than I thought at the time.

Learning to cope with rejection and knowing whether to modify or abandon a severely criticized approach are probably the most useful skills acquired during my long career. Looking back at the two-neutrino experiment it is remarkable that the 1962 paper reporting its results listed only seven physicists as authors. In contrast, a major experiment in particle physics today might have a hundred or even a thousand times as many authors. For example, the two teams that discovered the Higgs particle in 2013 included 3000 scientists on one team and even more on the other. In some cases, the list of authors is longer than the article itself!

Good and Evil Tachyons

If the Higgs boson has been called the *God Particle* (the title of a book by Leon Lederman[8]), perhaps I may be forgiven for referring to tachyons as coming in "good" or "evil" varieties – although I do worry this terminology may be taken too literally by some readers. In their good variety, as in the case of the famous Higgs particle, an imaginary mass tachyon is claimed to create an instability in the so-called Higgs field. After the excitations created by the instability spontaneously subside, there are no observable particles with either FTL speed or imaginary mass. When most theorists refer to tachyons, they are talking about this "good" variety. For example, of the 4360 papers published since 2017 that contained the word *tachyon,* only 12% also contained the word *superluminal,* and only 4% contained the word *neutrino* as well. Directly observable particles having superluminal (or FTL) speed or imaginary mass would be the "evil" ones, because they would violate some core physical principles currently *assumed* to be correct. Some theorists have even claimed that the existence of such tachyons would spell doom for the whole universe because they could be created from the vacuum of space in pairs having equal magnitude positive and negative energy, thereby making the whole universe unstable. Not surprisingly, most theoretical physicists have had little use for the FTL tachyon. In the remainder of this book the word tachyon refers to these "evil" ones, but I may sometimes add "FTL" for emphasis.

Seeing Things

I do not consider myself much of a theorist, since I lack the needed deep mathematical knowledge required to be a good one, nor do I consider myself much of an experimentalist either, since I was never very adept with hardware. Rather, my main contributions in physics have been in the realm of data analysis, and especially as a data prospector who is always seeking interesting patterns in data that have been overlooked. Searching for overlooked patterns in data can be risky, since often they only exist in the eye of the beholder, which is what is normally meant by the colloquial phrase "seeing things." Two such examples from astronomy would include the infamous "canals" on Mars, and more recently the "face" on that planet's surface, with both claimed by some Earthlings, but no known Martians, to be evidence of a Martian civilization.

Interestingly, the myth of Martian canals was, in part, the result of a linguistic fluke since the Italian astronomer Giovanni Schiaparelli used the Italian word *canali* meaning "channels" to describe the features he observed in 1877, but the word was mistranslated into *canals.* These features *drawn* by Schiaparelli were taken as evidence by some scientists for Martian "canal designers," but they were essentially an illusion coupled with wishful thinking in an era before telescopic images were photographed. Despite the illusory Martian canals, known features on the planet's surface that, while not created by Martians, are believed to have been made by flowing water. Even more interesting, while it had long been thought there is no longer water on Mars, in 2018 NASA announced it had discovered evidence of a subsurface lake under the south polar Martian ice cap. Once considered to be rare except on Earth, water is now believed to be prevalent on many bodies in the solar system, with Earth having a mere 2–4% of the total amount. This observation makes the possibility of life in the solar system much more likely than previously thought. "European life," that is, life on Jupiter's moon Europa, seems like a particularly good bet given there is more water there than in all the oceans on Earth – seriously!

In contrast to the canals, the face on Mars (see Figure 1.8) has been championed by far fewer legitimate scientists than Schiaparelli's canals ever were.

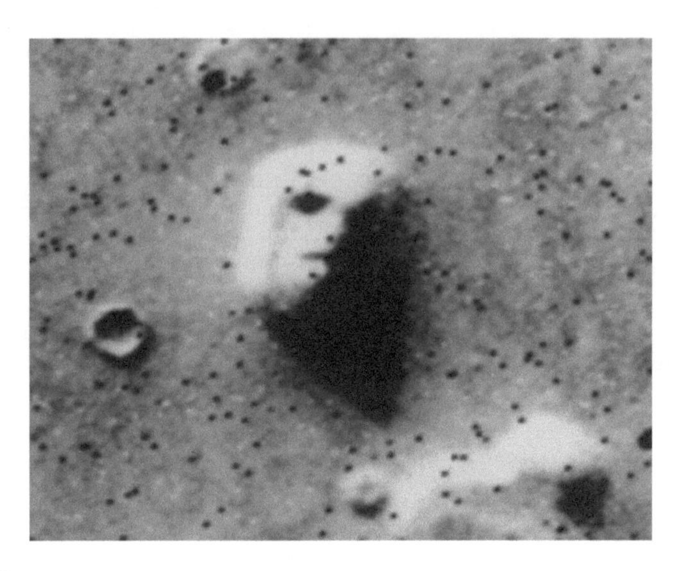

FIGURE 1.8
The "face" on Mars recorded during the NASA Viking missions to the red planet. Pub. Domain.

According to the main proponent of the face's artificial origin, astronomer Tom Van Flandern, there is only a 1% chance of it having a natural origin. Of course, humans are virtually programmed to see faces wherever they look, so this estimate cannot be taken at "face value." Calculating the probability that a random pattern resembling a face will be seen somewhere in all the areas of the Martian surface, all the clouds, all the pizzas, or all the spilled milk one sees on the pavement can be very tricky. Based on an analysis of higher resolution images NASA has stated: "A detailed analysis of multiple images of this feature reveals simply a Martian hill whose illusory face-like appearance depends on the viewing angle and [the] angle of illumination."[9]

The example of the face on Mars illustrates the danger of trying to find the odds of a pattern being due to chance only *after* it has been observed – a statistical pitfall known as a *post hoc* analysis. It is, for example, quite another matter if you can specify the nature or the location of the pattern you expect to see *before* examining a set of data (or before seeing the face on Mars). To take a crazy example, imagine that prior to seeing the face a radio transmission was received from the direction of Mars that said: "Hey Earthlings, look at something neat at 40.75° north latitude and 9.46° west longitude on the Martian surface." Such a message would drastically change our estimation about whether the face we then observed at that location had a natural origin!

A more plausible example of the *post hoc* analysis pitfall would be the case of "cancer clusters," which are geographic regions that have higher rates of the disease than the nation generally. Although cancer risk may well be elevated due to environmental influences, finding an elevated cancer rate in some area can easily be due to the tendency of randomly distributed data to cluster in space or time more often than we intuitively expect. To avoid the *post hoc* analysis pitfall, we would need to predict the areas of higher cancer rates ahead of time, say based on the location of power lines or cell phone towers. If cancer clusters were then found highly correlated with the position of power lines, that could be highly significant, but in fact, no statistically robust connection has been found in studies aside from one that was done in 1979, which was contradicted by later studies.

Returning to particle physics, I must confess having fallen into the *post hoc* analysis trap of seeing a pattern that was not there on more than one occasion. One memorable case was my claim that time comes in discrete chunks or quanta that was later proven to be invalid when more data on particle lifetimes became available. Despite such past errors, I have continued to believe that there are hidden patterns in particle physics remaining to be discovered, and that I might be lucky enough to find one. Most satisfying of all would be discovering something that most physicists have dismissed as either impossible or so outlandish as not worth their while to pursue. Best of all would be to have stumbled upon something (like the reality of FTL particles) that Einstein himself dismissed as impossible. Such fundamental discoveries, of course, are made only very rarely, since most "crazy" ideas turn out to be blind alleys. Thus, seeking overlooked patterns in existing data will likely show them to be either statistical flukes, or the result of some effect you overlooked, coupled with wishful thinking. On the other hand, if you find the same pattern occurs repeatedly in many data sets, maybe it deserves to be taken seriously. This sort of convergence of evidence from many areas (known as *consilience*) has led me to the conviction that one type of neutrino is very likely a tachyon.

Black Swans

Those few crazy ideas that do lead to new fundamental discoveries are the "black swans" of science, a phrase used by Nassim Nicholas Taleb in his 2007 book *The Black Swan: The Impact of the Highly Improbable*.[10] The term black swan arises because an observation of a single black swan (found mainly in Australia) after repeated observations of only white ones comes as a great shock to us. For many non-Australians, every time we see a white swan the belief that all swans must be white becomes *unjustifiably* stronger. The black swan phenomenon illustrates the problem of learning about the universe from induction, usually defined as the derivation of general principles from specific observations. Taleb defines a black swan event not only as one which most people consider extremely unlikely before it occurs but also one which when it does occur has important consequences. According to Taleb, black swans occur not only in science but also in history, economics, and politics. People are continually surprised when they occur, because they are falsely thought to be much rarer than they really are, based on models which entail incorrect assumptions. One example of a false assumption would be that the probability of rare economic events, such as market crashes, follows the familiar bell-shaped Gaussian distribution. Large crashes, however, occur more often than that because of the way panics develop.

According to Taleb, history is not driven by a smooth sequence of trends over time, but it is buffeted by black swan events that continually catch us by surprise, given their unexpected dramatic occurrence. Examples would include the onset of World War I, the 9-11 attacks on the World Trade Center, the rise of Donald Trump, and the Coronavirus pandemic. After the fact, most black swan events appear quite predictable given precursor events, as is the case for most of the five examples. The one exception is probably the start of World War I, arising from the assassination of an obscure Archduke in Serbia. In fact, even now, consensus among historians on the true origins of the Great War remains elusive.

At the beginning of this chapter, a different zoologic metaphor described unlikely occurrences, namely the unicorn, specifically in referring to the tachyon. The difference between black swans and unicorns is that the latter have been ruled out as being not merely unlikely but very likely non-existent or mythical. In popular culture unicorns have additional connotations. Sometimes they are assumed to be very well-hidden or shy creatures (as in some children's books and games), or else they refer to something that is extremely rare, and magical (as in a business start-up valued at more than a billion dollars).

The Neutrino as a Unicorn

The neutrino was first proposed by Wolfgang Pauli in 1930, and later given its name, meaning "little neutral one" in Italian by Enrico Fermi. Pauli and other physicists had been perplexed by the process known as beta decay in which an atomic nucleus emits a so-called beta particle, now identified as an electron. This process can be written as $X \rightarrow Y + e$, where X and Y are the so-called "parent" and "daughter" nuclei, and e is the electron emitted from the parent nucleus. If there were no other particles emitted besides the electron, one would expect it always to have some fixed energy E. That energy could easily be found based on the magnitude of the mass difference, m, between the X and Y nuclei using Einstein's $E = mc^2$. Instead of a fixed energy, electrons were observed to be emitted with a continuous range of possible energies – see Figure 1.9. It was as if some fraction of the available energy for the electron was

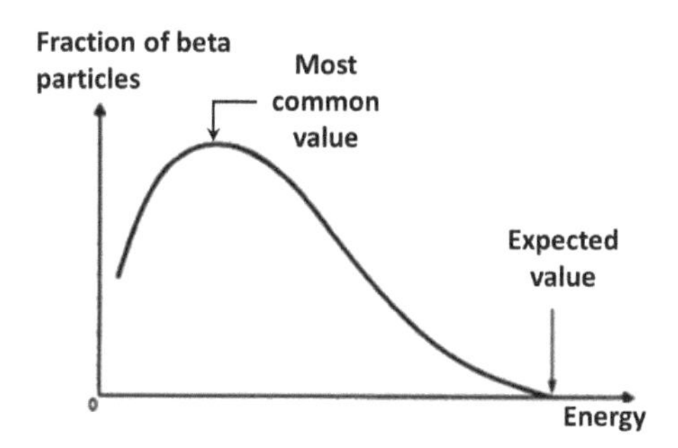

FIGURE 1.9
Spectrum of beta particles (electrons) in beta decay. "Expected value" assumes no neutrinos were emitted.

always missing – sometimes more and sometimes less. So serious was the "missing energy problem" that some physicists were even considering giving up on the sacrosanct law of energy conservation. Pauli's solution was to postulate the existence of a neutral particle (the neutrino) that was emitted along with the electron, but which escaped unnoticed with some fraction of the available energy.

Pauli's suggestion of the hypothetical neutrino came at a time when all subatomic particles were thought to be either positively or negatively charged, making the idea of a particle that was electrically neutral virtually unthinkable to most physicists – perhaps as unthinkable as tachyons today. When Pauli first made his proposal, he thought this neutral particle was in the nucleus before the decay. However, Enrico Fermi later realized that both the emitted neutrino and electron were both *created* during the decay process. Pauli considered his neutrino proposal, made in a letter to participants in a conference, so speculative as to preclude its publication in a peer-reviewed article. He even apologized for his suggestion by noting that "I have done something very bad today by proposing a particle that cannot be detected; it is something no theorist should ever do."[11] Pauli turned out to have underestimated the ingenuity of experimentalists here, but as a measure of the neutrino detection difficulty, it was another 26 years before the particle was observed by Clyde Cowan and Frederick Reines. Since that experiment in which neutrinos, or actually their antiparticles, the antineutrinos, were first detected, much has been learned about these unicorns, with a second type or "muon flavor" discovered in the two-neutrino experiment, and finally a third (tau) flavor a decade later.

Strangely, it has been found that those three neutrino flavors can morph into one another in the process known as neutrino oscillations. So, if you produce a beam of neutrinos having the electron flavor, they will spontaneously change into one of the other flavors after traveling some distance, and then further on change back into the electron flavor. Sometimes people claim that the existence of neutrino oscillations illustrates the weirdness of quantum mechanics, because it would be analogous to having a bunch of animals oscillate between being dogs at one time and then later being cats, and then dogs again, etc. However, while quantum mechanics is extremely weird, it is more correct to say the existence of oscillations simply shows that the flavor of neutrinos is not an immutable property as was originally thought. Thus, rather than use the cat-dog analogy think of the oscillating neutrinos as akin to a change that could in fact occur, such as between male and female west African frogs, which spontaneously switch their sexes. This change likely occurs when the population does not have enough of one sex to allow effective procreation.

Neutrinos have revealed their secrets only very slowly and have been the most puzzling and surprising of all the elementary particles. Progress has been slow because neutrinos are very difficult to detect given their extremely weak interaction with other matter. As a result, neutrino experiments require either large detectors, intense sources, or a long observation time, if we seek to detect a significant number of them. It should therefore not be too surprising that discoveries relating to neutrinos have been one black swan event after another.

THE LONG HISTORY OF NEUTRINO SURPRISES

- The missing energy in beta decay led Pauli to suggest them contrary to the belief that all atomic constituents were positively or negatively charged.
- Their detection 26 years later occurred contrary to Pauli's belief that they were virtually undetectable.
- A 70% deficit was observed in the number of neutrinos from the sun reaching Earth, contrary to the standard model of solar output.
- A second and third neutrino "flavor" were observed, contrary to expectations.
- Neutrino oscillations were observed showing that some have nonzero mass, contrary to the standard model that predicts they are massless.
- Neutrinos were found to be left-handed contrary to all other particles which have no specific handedness.
- Contrary to the belief that right-handed neutrinos do not exist, there are strong indications they do. Such "sterile" (inert) neutrinos would feel only gravitational force, unlike the active neutrinos that feel the so-called weak force as well.

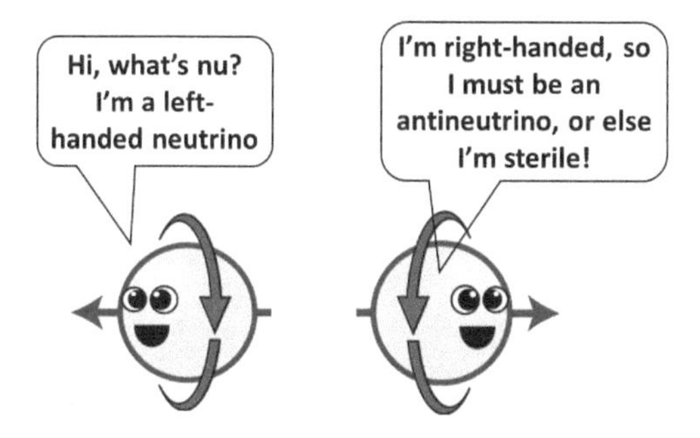

FIGURE 1.10
Neutrinos and antineutrinos have a definite handedness based on the relative spin and motion directions.

One of the strangest of the listed neutrino surprises above may be the idea that they are strictly left-handed particles, unlike all other particles, which have no specific handedness. In order to understand handedness, consider the quantity known as spin possessed by neutrinos and many other subatomic particles. While particles having spin may behave like spinning tops in some ways, they are not literally rotating contrary to the cartoon in Figure 1.10. In any case, among particles with spin, only the neutrinos have a handedness, that is, a fixed relation between their spin axis and their direction of their motion. With neutrinos the relation between their motion direction and that of their spin direction is like that between your thumb and curled fingers of your *left* hand, while for a neutrino's antiparticle (the antineutrino), the relationship is like that of your right hand – see Figure 1.10. Right-handed neutrinos and left-handed antineutrinos simply do not exist in nature, unless they are of the "sterile" variety. Sterile, that is, inert neutrinos would feel only the force of gravity, and not the so-called weak force, which is actually far stronger than gravity. The preceding discussion, however, assumes neutrinos and antineutrinos really are different particles, and not merely right- and left-handed versions of the same particle – an issue that is currently unresolved.

Models of the Three Neutrino Masses

Throughout this chapter the terms *models* and *theories* have been used, so it might be useful to clarify their difference. A model is an evidence-based representation of something that is either only an approximation or else too difficult to display. In contrast, a theory is a comprehensive explanation for patterns in nature that is supported by evidence. A wonderful saying by British statistician George Box conveys the need for caution when evaluating models: "All models are wrong, but some are useful."[12] The most

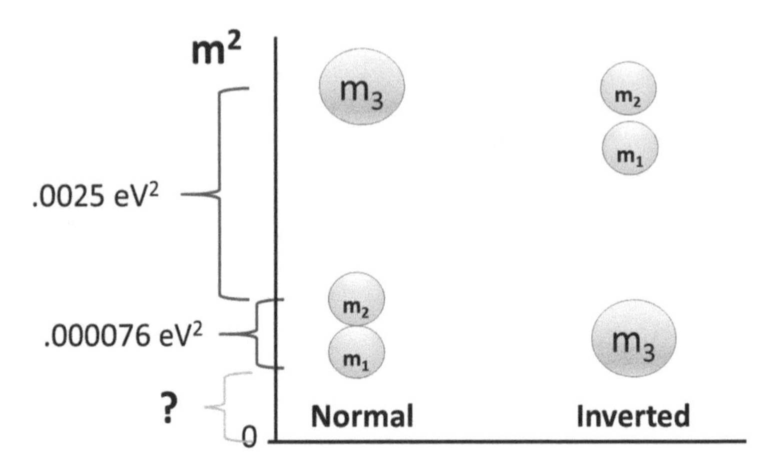

FIGURE 1.11
Two variations of the standard model of three neutrino masses.

useful models are those which are closest to reality, involve well-understood approximations, and make testable predictions. Sometimes even a "spherical cow" could be a reasonable and useful approximation. That ridiculous phrase is from a joke that spoofs the simplifying assumptions that are often used by theoretical physicists. In the joke, a brilliant physicist is asked by a farmer how his cows can be made to yield more milk and she begins by saying: "First consider a herd of spherical cows that can breed in vacuum."

Later we will be considering an unconventional model of the neutrino masses that includes tachyons, so let's first describe the commonly accepted model in which there are only three neutrinos having unknown (but nearly equal) masses m_1, m_2, and m_3. The three flavors of neutrinos (electron, muon, and tau) are each composed of a different combination or mixture of those three masses, but there are two variations of the model depicted in Figure 1.11 known as *normal* and *inverted*. Although the masses of the three neutrinos are unknown, there are known separations between the squares of the masses of each pair of them (see figure) based solely on the measured wavelengths found in neutrino oscillation experiments. If any sterile neutrinos are present in the standard neutrino model their masses are assumed to be extremely large compared to those of the three active neutrinos. The key points to remember about this conventional model are that there are assumed to be: (a) only three neutrinos with (b) very closely spaced, but unknown masses, with (c) none having $m^2 < 0$. Most physicists accept this model, and they believe it to be so firmly established that they don't even refer to it as a model but rather the "neutrino mass hierarchy."

The rest of this chapter addresses some issues concerning tachyons needing a bit further elaboration, namely,

- **Other tachyon possibilities**. Why not some particle other than neutrinos?

- **Ockham's razor**. Why not choose the simpler possibility, i.e., no tachyons?
- **Sending messages back in time**. Is this possibility even conceivable?

Other Tachyon Possibilities

Earlier it was noted that among all the known particles only one or more of the neutrinos might be tachyons. One reason is because their masses squared (m^2) are so close to zero, experiments have not yet revealed if m^2 is positive or negative. In addition, only neutrinos have never been measured to definitively have a speed *less than* light. But why need tachyons be one of the known particles? Why could they not be some particle not yet observed, or one that does not interact with other matter? The latter case would be of no interest to most physicists since no observational test would be possible. However, one could imagine their interaction with other matter was only through gravity (just like sterile neutrinos), which might allow observation.

Another alternative is that tachyons exist, but their properties are such that they could only be produced in accelerators having an energy above those now existing. Even if current particle accelerators lack the energy needed to create tachyons, one could still look for them in the cosmic rays that bombard the Earth from space, which can have energies well above any accelerator. One might recognize that a given cosmic ray particle was a tachyon in multiple ways, the simplest is to time their flight over some known distance to measure their speed.

In fact, two researchers, Roger Clay and Phillip Crouch in 1974 once claimed to find possible evidence that some cosmic ray particles were tachyons. However, their results were not replicable, and the nature of their speed measurement was very indirect. Still, one cannot rule out the possibility of an "out-of-the-blue" surprise in the future, but the idea of some new hitherto unobserved particle being a tachyon is much less interesting than that of one of the neutrinos being a tachyon – the principal subject of this book. The reason is three-fold. First, much theoretical work has already been done about neutrinos being tachyons. Second, there are many empirical reasons supporting this possibility, discussed at length in subsequent chapters. Finally, with no theory suggesting some unknown "T-particles" exist having any specific mass (or other properties), and no empirical evidence suggesting their existence either, it is impossible to know where and how to look for them. In contrast, with tachyonic neutrinos, a model has been put forward, which should be tested by an experiment now underway (known as KATRIN), with a definitive resolution possible in the coming few years.

Using Ockham's Razor Can Be Dangerous

Suppose we accept that some evidence exists favoring neutrinos as tachyons, but there is no decisive experiment for them yet, and many serious theoretical objections against them. "Ockham's razor" (named after English Franciscan friar William of Ockham, c. 1287–1347) tells us to choose the simpler possibility, that is, that tachyons do not exist. However, that inference would be an invalid use of Ockham's razor, because until an experiment shows that neutrinos are *not* tachyons, the possibility remains that they are. As previously noted, three ways to show that neutrinos are tachyons would be to find that: (1) their speed is below light speed, (2) their mass squared is negative, and (3) higher energy neutrinos reach us after lower energy ones. On the web site *Ehrlich.physics.gmu.edu* I list five additional ways to learn if neutrinos are tachyons. As of late 2021, no experiment has yet shown neutrinos are tachyons, but equally important, no experiment has shown they are *not*. Barring an unexpected supernova in our galaxy, or a positive sighting in the KATRIN experiment, this ambiguous situation is unlikely to change in the very near future given the difficulty of the relevant experiments.

In later chapters we shall consider various sorts of circumstantial evidence that suggests some neutrinos are tachyons. Some of this evidence might be a coincidence, but I believe the totality of the evidence is unlikely to have occurred by chance, assuming neutrinos are *not* tachyons. While I know the face on Mars was strictly in the eye of some beholders, but after discovering this evidence for neutrinos being tachyons, it was almost as if I observed the surface of Mars with a high-resolution telescope and seen not a crude face but instead something much more improbable – see Figure 1.12. OK, I know this analogy is ridiculous but you get the idea.

Sending Messages Back in Time

It is far from certain that tachyons could send messages back in time even if they are proven to exist. However, the mere possibility of this occurring has been one reason some physicists have doubted their existence. Obviously if tachyons could be used to send messages to the past, this would allow future scientists to send information to the present. The alleged ability to receive information from the future is usually considered to be an extrasensory "psychic" ability known as precognition. Clearly, if precognition existed, either as a sensory or extrasensory type, it could not be an everyday phenomenon for many people, otherwise there would be far more multiple lottery winners and casinos going broke.

FIGURE 1.12
"Alien Vacation: Mount Rushmore," by Mike McGlothlen, http://mike-mcglothlen.pixels.com

There are at least seven non-supernatural explanations for the extreme rarity of precognition, each having to do with the nature of the message, the technology or the person receiving it – maybe you can think of some more?

1. The phenomenon of receiving messages from the future does not exist.
2. Only certain gifted people can receive such messages.
3. Only in extreme cases, e.g., a premonition of disaster, is future information accessible.
4. Information "leakage" from the future occurs all the time but the signal is hidden by the much larger noise or "background."
5. The technology for receiving messages from the future does not yet exist.
6. The technology exists but we are unaware of it or fail to interpret the transmitted message.
7. It occurs only in dreams.

Options 4, 5, and 6 above would not be "psychic" abilities since they would not rely on extra-sensory means. Throughout my life I have been skeptical about precognition, and have strongly leaned toward possibility 1, while not being willing to rule out some of the others, provided compelling evidence was available. Of the remaining possibilities, I have been particularly skeptical about number 2, based on the actual track record of those who claim to have a psychic or precognitive ability. According to a fact sheet prepared by Rory Coker, Professor of Physics at the University of Texas at Austin, these "seers" invariably manage to rewrite their prophecies in retrospect so that they conform better with events. Nevertheless, despite most self-proclaimed seers being fakes or perhaps sincere but deluded, conceivably a handful of people may have seen the future clearly. My candidates would include perhaps Leonardo Da Vinci, H. G. Wells, and Jules Verne. Although I have had a few déjà vu experiences, I have always attributed these to coincidences, whose impossible-to-calculate probability cannot really be considered evidence for any precognitive ability. Of course, one sufficiently striking incident might make even this skeptic a believer. I suspect this might well be the case if I chose not to board a plane after a premonition of disaster, and the plane then crashed – particularly since I have never had any prior premonition.

Among the seven listed possibilities, number 4 is the one best suited to an empirical test, and in fact, these tests have been conducted. In one series of nine experiments involving over 1000 participants Daryl Bem of Cornell University looked for a response by subjects to a visual stimulus, but he tested for retroactive influence by "time-reversing" the stimulus and response, i.e., he looked for an individual's responses *before* the stimulus events occur. Bem claimed that statistically significant effects were seen in all but one of the nine experiments he conducted. His 2011 article has been very widely cited (713 times), and it has been called the most convincing parapsychology study dealing with precognition ever conducted. Unfortunately for Bem, however, when other scientists repeated his experiments no such effect was seen. In addition, various flaws have been found in his work including using a very low standard of statistical significance and his changing his methods part way through the experiments. Most damningly, in 2016, Bem and two coauthors reported the results of a replication experiment that was conducted using more rigorous methods, and no evidence of precognition was found. Sadly, these negative results are cited far less often than Bem's original positive study. In summary, no convincing empirical evidence exists for precognition, despite claims to the contrary on the part of believers. Perhaps if there is any possibility of receiving messages from the future, it falls in category 5 or 6, with tachyons a prime candidate for transmitting them, a possibility recognized a century ago by Einstein and others shortly after relativity was discovered, which is one reason many physicists have ruled out the existence of FTL particles.

"Not so fast," say four theorists Heinrich Pas, James Dent, Sandip Pakvasa, and Thomas Weiler. In their 2006 paper in *the Physical Review* these researchers describe a form of string theory with extra dimensions beyond the usual four of space and time. In their theory two extra dimensions are assumed to be asymmetrically warped, in such a way that the speed of light varies with position. Their theory allows for shortcuts through the extra dimensions which could be traversed by sterile neutrinos. Pas and his colleagues show that these shortcuts would allow the neutrinos to have FTL speed and travel back in time. Thomas Weiler of Vanderbilt University, one of the paper's authors, has gone so far as to suggest that if their theory is right, and if sterile neutrinos exist, it might be worthwhile to use a giant neutrino detector to look for messages from the future.

My Message from the Future

I would not be so foolish as to try to build a sterile neutrino detector myself as Weiler suggested to look for messages from the future, but in recent years I have been working on an unconventional neutrino mass model having three sterile neutrinos, including one that is a tachyon. While I was engaged in this research, I had an extremely strange experience for which I have no logical explanation. On November 22, 2015, work on my 3 + 3 neutrino model was at a highly preliminary stage, but I had foolishly submitted an article about it to *the Physical Review Letters* – premature submission being a real danger when one works on a research project by oneself. Although the article was deservedly rejected, a greatly improved version was eventually published in the journal *Astroparticle Physics* in 2018. Aside from the grossly incorrect content, I had initially chosen a terrible title for the article, "Evidence for a tachyonic mass-eigenstate neutrino having $m^2 \approx -0.2$ keV^2, based on certain features of data taken on the day of SN 1987 A in two detectors." Most journals nowadays have online submission systems. As soon as I entered the atrocious title for my article into *the Physical Review Letters* manuscript submission system, I noticed that a greatly improved title was listed immediately after the title I had typed:

> Mont Blanc mystery solved: An $m^2 = -0.2$ keV^2 neutrino.

This startling event led me to marvel at the online submission system of the journal, which enabled it to suggest an improved title, and to do so with virtually no time delay. A day after submitting the article I thought more about the online submission system's ability to suggest improved titles, and I could not imagine how it might conceivably work. I therefore contacted the Interim Editor in Chief and Editorial Director, Dr. Daniel T. Kulp by email to learn more about their system. Dr. Kulp assured me that the journal had no such system to suggest improved titles. After recovering from my shock, I could imagine only four explanations for my experience the day before. The simplest possibility is that I imagined the improved title on the screen. While I have sometimes misread messages in emails, this was such a surprising event when it happened that I re-read what appeared on the screen many times, and I am quite certain I did not misread what I saw. A second possibility is that Dr. Kulp had been mistaken about no such software for suggesting improved titles existing on their manuscript submission system. He was after all "Interim" Editor in 2015, suggesting that he was new on the job. This possibility was foreclosed after I contacted Dr. Kulp again in September 2018, and he confirmed his earlier assertion.

Here is a third conceivable explanation for my experience. Let us suppose that tachyons really could be used to send messages back in time. Perhaps someone in the future had sent me this message as a form of encouragement to keep at it, because I was on the right track. If so, they probably would not just send me an email offering encouragement, and claim to be from the future, knowing I would just dismiss the email as a joke or scam. The "message" was much more effective in terms of its impact on me because of its mysterious origin. This explanation may strike the reader as far too implausible to be true, but whatever its origin, since the event occurred at a point when my model was at a very confused and preliminary state, the "message" did give me a great deal of hope that I was on the right track. I have no idea what technology might make such a thing possible, and especially, how a message from the future using tachyons could interact with the web-based manuscript submission system of a journal. Of course, it must also be recognized that a message from the distant future might rely on technology that would seem magical to us today. As scientist and science fiction writer Arthur C. Clarke has said: "Any sufficiently advanced technology is indistinguishable from magic."[8] While we are considering truly bizarre explanations for my "message" here is an even more bizarre fourth one. Perhaps I am not the original Robert Ehrlich who by luck stumbled upon an important discovery. Instead, I am a consciousness living in a virtual reality simulation of the original Ehrlich's life being run by humanity's super-advanced descendants. I'm not sure what that makes you dear reader. Stay tuned for more on this crazy idea in the final chapter.

Summary

Albert Einstein explicitly ruled out the possibility of FTL particles when he showed that an infinite energy would be needed to accelerate a particle up to or beyond light speed in 1905 in his first of two papers on special relativity. Nevertheless, in 1962 George Sudarshan and two colleagues, O.M.P Bilaniuk and V. K. Deshpande, showed in their "Meta" relativity paper how hypothetical particles, now known as tachyons, might have FTL speed if they had an imaginary mass and they always moved at superluminal speed. These hypothetical particles could then be instantly created in subatomic collisions with FTL speed without requiring an infinite energy. Still, most physicists have been dubious about the existence of FTL particles, both on theoretical and empirical grounds, including the lack of any conclusive experiment proving their existence.

Among all the known particles, neutrinos are the only candidates for being tachyons, because (a) their mass is so close to zero that it is unknown whether the mass is real or imaginary (or whether m^2 is positive or negative), and (b) no neutrinos have never been *surely* observed traveling slower than light. Although most physicists are dubious about FTL tachyons, a minority believe there is evidence that they do exist, and specifically that some neutrinos are tachyons. The main thesis of this book is that tachyons are not mythical creatures but rather they have had very good camouflage. Seeing through that camouflage has involved a method called data prospecting, which involves examining published data in novel ways, as explained in subsequent chapters. A final part of this first chapter discussed the possibility that tachyons might be able to send a message back in time, which raises some of the same paradoxes as time travel. The backward time traveling aspect of tachyons, and how that property can be translated into a method for detecting them, is the main subject of Chapter 2.

References

1. Galchen, Riva. quoted in Dan Ariely, "The Best American Science and Nature Writing," New York, NY, Houghton Mifflin Harcourt, p. 259, 2012.
2. Bilaniuk, O. M. P., V. K. Deshpande, and E. C. G. Sudarshan. "Meta" Relativity, American Journal of Physics, 30, no.10, 718–723, 1962.
3. The Yale Book of Quotations by Fred R. Shapiro, Section: Arthur Buller, Page 113, Yale University Press, New Haven, 2006.
4. The Lincoln Evening Journal (Lincoln Journal Star), Little energy from atom (Associated Press), Page 2, Column 2, Lincoln, Nebraska, September 11, 1933.
5. As quote appeared in the Journal of Advertising Research (March-April 1998).
6. Pais, Abraham. Inward Bound, Oxford, Oxford University Press, p. 418, 1986.
7. Clarke, Arthur. C., Clarke's Third Law on UFO's, Science, 159, no. 3812, 255, 1968.
8. Lederman, Leon, and Dick Teresi. The God Particle: If the Universe Is the Answer, What Is the Question? New York, NY, Dell Publishing, 1993.
9. "The Face on Mars." Image of the Day Gallery. NASA. Retrieved April 26, 2007.
10. Taleb, Nassim Nicholas. The Black Swan: The Impact of the Highly Improbable, New York, NY, Random House, 2007
11. As quoted in APS News, July 2011, Volume 20, number 7.
12. Box, George E. P., *Science and Statistics*, Journal of the American Statistical Association, 71, no. 356, 791–799, 1976.

2

Faster than Light and Backwards in Time

... the distinction between past, present, and future is only a stubbornly persistent illusion.[1]

Albert Einstein

Tachyons in Fact and Fiction

The idea of sending messages back to an earlier time using tachyons has been the theme of movies, short stories, and novels, including the 1980 award-winning classic "Timescape,"[2] by science fiction writer and astrophysicist Gregory Benford, who has also published a scientific article on tachyons. My first foray into tachyon research was similarly connected to a science fiction story I had written, although mine was never published. In 1997 *Physics Today* magazine ran a contest inviting readers to submit a piece about an imagined future discovery in physics. My entry was about an international team that had succeeded in 2050 in using a beam of tachyons to send a message back to our own time. My story was made even more far-fetched by having the effort led by an Iranian-Israeli duo, but I was not the message recipient in this case!

While my entry did not win, the exercise stimulated my interest in tachyons and led me to consider ways that evidence for them might be found. I had been aware that neutrinos were the only tachyon candidates among the known particles, and I began to imagine how one might observe backward-time-traveling neutrinos. The idea that faster than light (FTL) particles can go backward in time does seem to defy common sense. For example, you would think a tachyon signal emitted from Earth toward a distant receiver would arrive later than it is sent, no matter how high is its speed. However, for tachyons a weird switch can occur between the identity of the sender and receiver of the signal, which is what is meant by reversing its direction in time. The first part of this chapter will explain how this works, based on the math of Einstein's Special Theory of Relativity. Fortunately, that math can be explained with pictures and imagined experiments rather than equations. Einstein himself often used such "Gedankenexperiments" (German for thought experiments), to help him clarify his thoughts.

DOI: 10.1201/9781003152965-2

A Lazy Dog That Could Not Find an Academic Job

Albert Einstein was an obscure 26-year old technical expert third class in the Swiss patent office when he published his first relativity paper in 1905. Among those who quickly came to appreciate the significance of that paper was Hermann Minkowski, his former mathematics professor. Minkowski noted that Einstein's achievement "came as a tremendous surprise, for in his student days he had been a lazy dog ... He never bothered about mathematics at all."[3] Note, however, that the mathematics behind special relativity involves little beyond high school algebra and trigonometry. Einstein's great achievement in inventing special relativity was not primarily a mathematical one, but a new way of looking at space and time. His first relativity paper is remarkable in several respects. In addition to having a very obscure title: "On the electrodynamics of moving bodies," it also had zero references, something virtually unheard of for scientific papers. In addition, there was no mention of the famous $E = mc^2$, which was the subject of a second relativity paper that same year. In fact, 1905 has been called Einstein's "miracle year," since he published four papers on very different subjects, three of which are among the greatest physics papers ever written. Of the four papers, C. P. Snow has written:

> The conclusions, the bizarre conclusions, emerge as though with the greatest of ease: the reasoning is unbreakable. It looks as though he had reached the conclusions by pure thought, unaided, without listening to the opinions of others. To a surprisingly large extent, that is precisely what he had done.[4]

Despite these miraculous achievements, five years earlier Einstein's prospects were less than promising. In 1900 Einstein struggled in vain to land an academic job after barely finishing his graduate studies at the Zurich Polytechnic Institute. His failure to land an academic job was probably because he had antagonized one too many of his professors by attending very few lectures and instead relied on lecture notes taken by his friend Marcel Grossmann to pass his exams. Unable to find a teaching job, Einstein finally secured employment, with Grossmann's assistance, at the Patent Office in Bern Switzerland. This was probably the best job in the world to prepare him to make his discovery of relativity. Einstein could complete a day's work in just a few hours, giving him a lot of free time to think. In addition, the nature of the work may have led him to a key insight behind relativity. At the Patent Office some of the proposals Einstein reviewed probably involved methods for synchronizing clocks in different locations, which was essential for enabling trains to follow railroad schedules. In fact, the Bern Patent Office was a clearinghouse for patents on the synchronization of clocks. Einstein was therefore led to think about how we know two clocks strike the hour at the same time when they are widely separated, perhaps in different cities.

As you are likely aware, time in relativity is assumed to be the fourth dimension, which when combined with the three dimensions of space, results in the notion of *spacetime.* When Professor Minkowski coined that word, he explained that: "Henceforth, space by itself, and time by itself, are doomed to fade away into mere shadows, and only a kind of union of the two will preserve an independent reality."[5]

The idea of time as the fourth dimension, however, was described much earlier than Minkowski. Two decades before relativity the concept was explained by the fictional time traveler in the H. G. Wells novella "The Time Machine."[6] Even more remarkably, in 1754 the French mathematician Jean d'Alembert mentioned the idea in an encyclopedia article, in which he wrote:

> A clever acquaintance of mine believes that one might nevertheless consider timespan as a fourth dimension, and that the product of time with volume would ... be a product of four dimensions; this idea may be contested, but it has ... merit, if only because of its novelty.[7]

Normally, when d'Alembert referred to ideas or works of others, he mentioned names as well as precise publications. Therefore, barring an interaction with H. G. Wells' time traveler, it is reasonable to assume that d'Alembert himself was his own "clever acquaintance."

The Block Universe and Its Worldlines

Perhaps the most important element of our perception of time is that of an ever-advancing present moment (a "now") separating past from future. However, Einstein, as well as many philosophers of science, have regarded the passage of time with its ever-advancing now as an illusion, albeit a very powerful one. Einstein's view of time is compatible with the block universe picture, according to which the universe is a giant four-dimensional block of all the things that ever happen at any time and at any place. In the block universe description, the past, present, and future are all equally real. Any event occurring at some point in spacetime can be defined by four coordinates x, y, z, and t, which specify where and when the event occurs. An *event* could be something as simple as a moving subatomic particle occupying a particular location at a particular time. Our universe is comprised of an incredible number subatomic particles – one estimate being 10^{86}. Most of these particles are neutrinos, about 100 trillion of which pass through your body each second.

Each of the particles in the block universe weaves a path through spacetime, known as its *worldline.* Even living beings are made up of an enormous number of particles, whose worldlines together form an intertwined bundle – see Figure 2.1. Of course, you are not made up of a fixed number of particles,

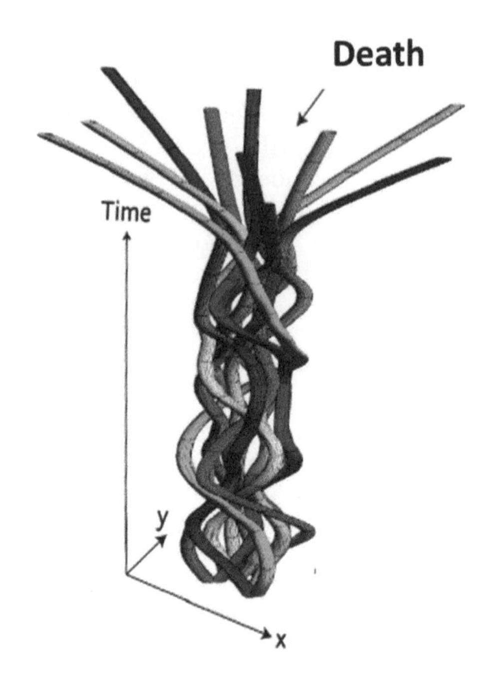

FIGURE 2.1
Worldlines for some of the particles making up a living organism.

Source: Image provided by Max Tegmark

because some are continually added and others removed, due to interactions between you and the environment. In fact, it has been estimated that only between 2 and 5% of the atoms making up your body at birth remain with you for your entire lifetime, and even more remarkably, about 98% of the atoms in your body are replaced every year! Upon your death and your body's subsequent disintegration, however, the worldlines of all the atoms of your body will tend to diverge as they go their own way, and your body returns to dust. In the block universe picture, the collection of 10^{86} worldlines for all the subatomic particles in the universe encompass its entire history, which Pastafarians might term our "spaghetti universe" – see Figure 2.1.

Challenges to the Block Universe Concept

Some philosophers reject the whole concept of the block universe. In the words of British philosopher John Lucas, " … it fails to account for the passage of time, the pre-eminence of the present, the directedness of time and the difference between the future and the past."[9] Of course, based on the quote opening this chapter, Einstein apparently considered most of these

Pastafarianism is a satirical religion whose deity is a Flying Spaghetti Monster. It began in 2005 when a 24-year old physics graduate, Bobby Henderson, wrote a letter to the Kansas Board of Education. Henderson had expressed opposition to the teaching in public school science classes of intelligent design as an alternative to evolution. Intelligent design is the theory that life cannot have arisen by chance and had to be designed and created by some intelligent entity. Henderson noted in his letter that Pastafarianism was just as valid as intelligent design, and he called for equal time for teaching this creed alongside both intelligent design and evolution in public schools, giving each equal time. Henderson explained: "I don't have a problem with religion. What I have a problem with is religion posing as science. If there is a God and he's intelligent, then I would guess he has a sense of humor."[8]

concepts to be "persistent illusions." Others who reject the block universe do so because they argue that consciousness, not matter, is the universe's main feature. The ultimate extension of this idea is that the entire universe is conscious, a notion referred to as *panpsychism*.

Since this is a book about FTL particles, it should be noted that some people claim that there are links between consciousness and tachyons. I have nothing positive to say about this topic, but the broader idea that the fundamental reality of our universe involves consciousness is not an absurd one. In fact, the connection between consciousness and quantum theories has been explored by such physics luminaries as Erwin Schrodinger, John von Neumann, and Eugene Wigner.

Worldlines and the Light Cone

Recall that all the subatomic particles making up the block universe form a collection of worldlines. Although a worldline describes the motion of a particle, just as its trajectory does, the two concepts are very different. A trajectory defines a particle's evolving path through space, while a worldline shows its path through space*time*. In other words, each point on a worldline shows when an object was at a specific location, which is unspecified by the shape of its trajectory.

Since spacetime is four-dimensional, we cannot directly depict it in a two-dimensional illustration. The best we can do is to ignore one of the three space dimensions, as in Figure 2.2, which shows two worldlines – one a helix and the other a vertical line. Can you imagine what sort of motion these worldlines might represent? The horizontal plane slices through spacetime at a fixed time, and we can imagine it to move steadily upward as time

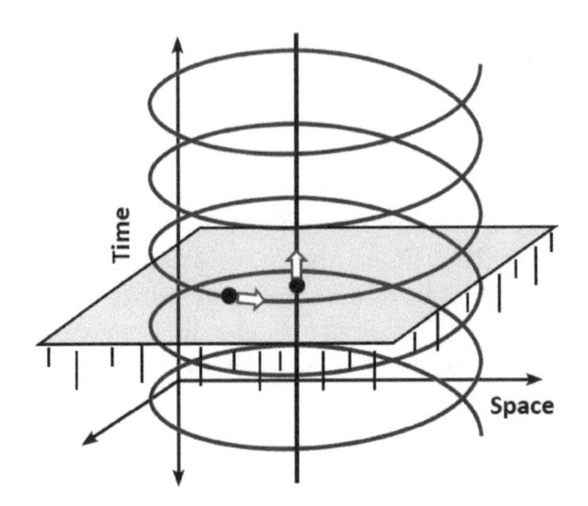

FIGURE 2.2
The helix is the worldline for a planet orbiting the sun whose worldline is vertical.

progresses. At the moment depicted, this horizontal "time plane" cuts the two worldlines at the two black dots shown. As time progresses one of the dots moves directly upward, or forward in time, while remaining stationary in space. The other dot orbits the first one in a circle, so the two worldlines might describe a planet orbiting a stationary sun.

Figure 2.3 shows another spacetime diagram that includes the so-called light cone, which consists of an upright ("future") cone and an inverted ("past") cone. Here the units used for time are years and those for space, light-years – a light-year being the distance light travels in one year. Given these units, the worldline for a particle moving at the speed of light is inclined at 45 degrees to the vertical time axis. Worldlines for a particle traveling at speeds less than light, and faster than light, are also depicted. The former

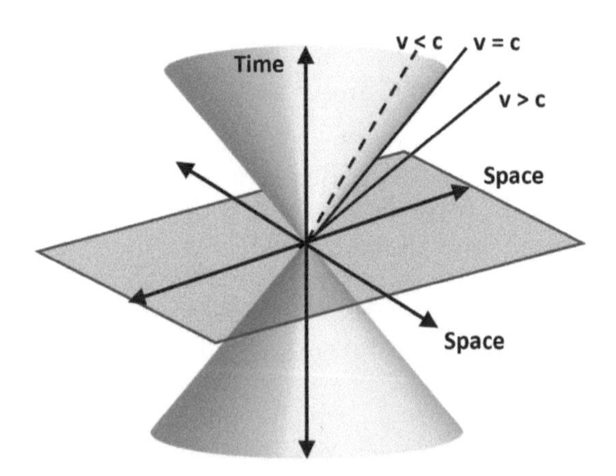

FIGURE 2.3
Spacetime diagram with three worldlines and the light cone.

worldlines are called "timelike," being closer to the vertical time axis, while the latter are called "spacelike," and they would describe tachyons.

The light cone in Figure 2.3 shows how a pulse of light would spread out in time in ever expanding circles as time advances. To get the idea, just imagine the horizontal plane shown in the figure rising upwards, or forward in time, thereby intersecting the upper (future) light cone in circles of larger and larger radius. Had that plane started rising from below its present position it would have intersected the inverted (past) light cone in smaller and smaller circles, eventually contracting to a point before expanding – something not normally seen for light waves, but theoretically possible with the right initial conditions!

The light cone divides spacetime into three distinct regions: the *absolute future* (inside the upper light cone), the *absolute past* (inside the lower light cone), and *elsewhere*. The reason for the "absolute" terminology here is that according to relativity, events falling outside the light cone (in the so-called elsewhere region) could be either in the future or the past, depending on an observer's state of motion. Such an inability to know for sure if an event is in the past or future conflicts with the common-sense view of time, and so let us see how it works according to relativity.

The Lorentz Transformation

This math of relativity is based on the Lorentz Transformation (LT), developed by Dutch physicist Hendrik Lorentz just prior to Einstein's relativity. It consists of a set of equations that can be found at many places on the web or in physics textbooks, but we only shall illustrate their *geometrical* form without any equations. Our goal is to show why the LT has the form that it does, and why tachyons can travel backwards in time as a result. Essentially, the LT describes how the space and time coordinates of events changes when you go from a "fixed" observer to a "moving" one – see Figure 2.4. The reason for the quotes here is that which of two observers is *really* moving cannot be decided – at least if we are dealing with constant velocities.

Notice that the figure uses normal rectangular coordinates for the stationary observer but not the moving one. For the moving observer, the space and time axes are each tilted away from the horizontal and vertical by an angle that depends on the observer's speed. Consider the worldline of a photon of light shown as a 45-degree line in Figure 2.4. As seen in the figure, this worldline describes the photon as having traveled 1 light-year in a time of 1 year to reach point P, according to the fixed observer. For the moving observer, who uses the slanted coordinates, the photon travels x light-years in x years to reach P. The photon's speed for the moving observer is distance divided by time, or x over x, which of course is again one light-year per year – the same speed as for the fixed observer. Having an identical speed of light

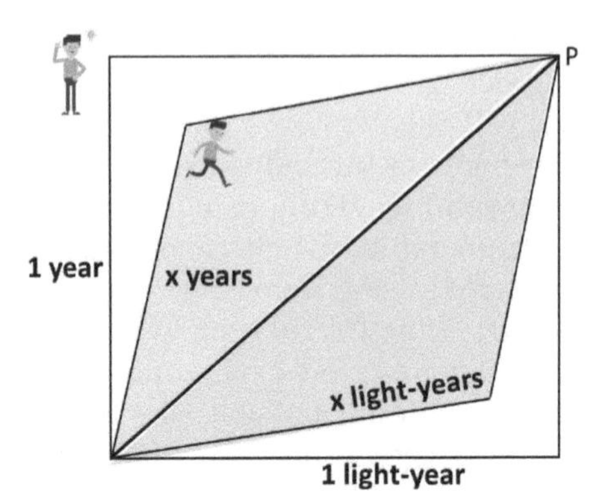

FIGURE 2.4
Geometrical example of the Lorentz Transformation relating the space and time coordinates of point P for fixed and moving observers.

for all observers is an essential aspect of relativity. In fact, Einstein started with this assumption of a constant speed of light, and he worked backwards to develop his relativity theory. Essentially what he did was to ask how the space and time coordinates had to change between two observers, in order to keep the speed of light unchanged, and he found that the LT did the trick.

Backward Time-Traveling Tachyons

Having seen how the coordinate system changes between two observers, based on the LT, let us now see why tachyons can change their direction in time, as a result. Figure 2.5 shows an imaginary gun firing a tachyon and a photon at the same instant. The tachyon worldline is depicted as the squiggly line, and that for the photon is shown dashed. As we would expect, the tachyon outraces the photon, since at a common "later" time it reaches point B which is further from the gun than the photon at point A.

Let us now see how things appear to the moving observer who uses the tilted coordinate system. For the moving observer, instead of showing the "later" time plane above the "now" plane, we instead show the "earlier" time plane below it. According to the moving observer, the dashed photon worldline still travels forward in time, and it still has the same speed c, as already noted. However, the tachyon worldline lies *below* the tilted "now plane" reaching point B′ at an *earlier* time than it left the gun! For the moving observer, we see that the tachyon has therefore traveled backward in time, since it reaches point B′ at an earlier time than when it left the gun. An equivalent interpretation of these events is that for the moving observer the tachyon went from point B′ to the gun, rather than from the gun to point B′. Now here is the

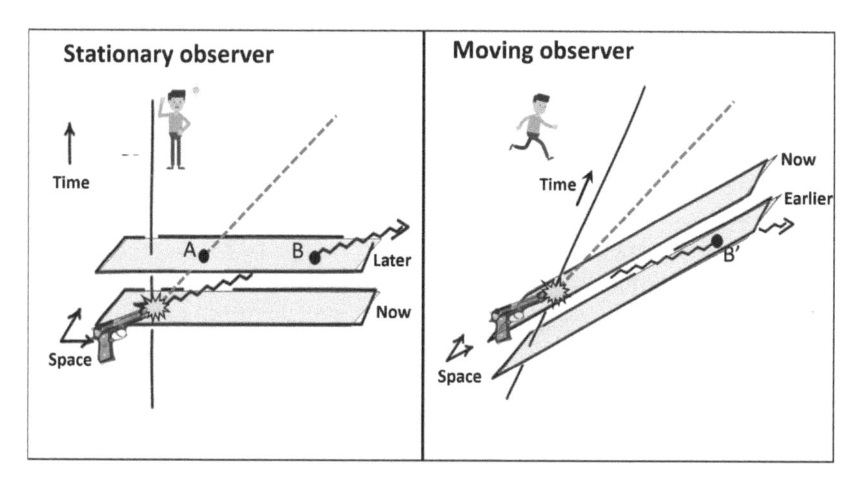

FIGURE 2.5
A tachyon changes its direction of time: the stationary observer sees it moving forward in time, but the moving observer sees it going backward.

important point: *according to relativity, we cannot say which direction the tachyon really went, or which of the two observers is correct.* The reason for this ambiguity is that both observers can regard themselves at rest and the other to be moving, so there is no preferred observer whose judgment we can trust.

The reversals of time-ordering that can occur for tachyons, cannot occur for ordinary particles traveling at sub-light speed. According to relativity, the time reversals that occur for tachyons are always accompanied by a switch in the sign of the particle's energy. Normally, we are not used to thinking of particles having negative energy of motion. However, if a negative energy particle is emitted from some interaction, we can apply a "reinterpretation principle," which equates an emitted negative energy particle to an absorbed one having positive energy – thereby reversing both the sign of the particle's energy and its direction in time.

Chasing a Tachyon

As a long-time tachyon hunter, I have been "chasing" tachyons, i.e., seeking evidence for them, for over three decades, but here I have in mind literally chasing one. Assuming no spaceship could travel at FTL speed, you could not, of course, catch up to a tachyon, but what would happen if you started chasing after one at a steadily increasing speed? This gedanken experiment is similar to one of Einstein's, who at age 16, wondered what he would see if he chased after a light beam. Here, however, the object of our imagined pursuit is a hypothetical tachyon not a photon. Suppose, for example, a

DO CAUSE AND EFFECT REALLY EXIST?

The tachyon gun example shows that the order of cause and effect can become reversed with tachyons. In fact, tachyons are said to violate the principle of causality. Thus, suppose the bullets from the tachyon gun were to kill someone at point B. In this case the "moving" observer in our example would see someone drop dead, and then the tachyon bullet would leave the corpse, and return to the gun from which it was fired. Despite the absurdity of this sequence of events, when describing the quantum world, we must abandon our everyday view of an absolute time sequence in which causes precede effects. Subatomic particles like electrons and tachyons (if they exist) are described by the rules of quantum physics. We usually describe quantum states starting with some initial state evolving forward in time, but the theory works just as well starting with a final state evolving backward in time to the present. In fact, in quantum physics one can have "mixed" (superposition state) in which one cannot say which of two events A and B is the cause and which is the effect. Thus, the absurdity of the preceding sequence with the bullets from a dead corpse returning to the tachyon gun is not a reason to maintain that tachyons cannot exist.

tachyon left Earth with a *constant* speed 437 times the speed of light, to take an arbitrary number. As you accelerated your ship chasing after it you would weirdly find that relative to your ship the tachyon sped away from you at an ever-increasing speed, almost as if it sensed your pursuit. In fact, once your ship speed reached 1/437 of the speed of light, the tachyon speed relative to your ship would approach infinity. This weird behavior is, of course, not due to any "tachyon consciousness" or its awareness of being chased! Rather, it follows directly from the equations of relativity, or specifically the "addition of velocities law." If your ship speed increased still further, and slightly exceeded 1/437th that of light, you would then find that the tachyon is now moving with near infinite speed but in the *reverse* direction. This reversal in tachyon motion direction, as judged by you, occurs just like the direction reversal with the earlier tachyon gun example when you, the observer, reach a certain speed. Could such a sudden reversal in the tachyon motion direction be observable assuming they exist? As we shall see, the answer is yes.

Searching for Tachyons

The reversal of a tachyon's direction of motion when it is chased at high enough speed is quite different from the reversal of motion which occurs when you chase an ordinary particle (or a car) and overtake it. Recall that the sign of a tachyon's energy also switches, and if a negative energy tachyon were emitted from some interaction, it could be reinterpreted as a positive energy absorbed tachyon. Therefore, it is more appropriate to think of these reversals as a switch in their direction of time, not as simple reversals in direction of motion. Energy is, of course, conserved in all particle reactions, but if a tachyon's energy can become negative for some rapidly moving observers, then we can show that some processes would be allowed even if they seem to violate energy conservation. The observation of such energetically forbidden processes would, therefore, be evidence that one or more of the particles involved in the reaction are tachyons.

These ideas were first clarified in the late 1990s in several papers written by Alan Chodos, and his collaborators Alan Kostelecky, Robertus Potting, and Evalyn Gates, who had proposed some experimental tests for neutrinos being tachyons. Their proposed tests involved beta decay, the simplest example of which involves the disintegration of a neutron into three particles: a proton, electron, and antineutrino:

$$\text{Neutron beta decay:} \quad \mathrm{n} \rightarrow \mathrm{p} + \mathrm{e}^{-} + \mathrm{v}.$$

The neutron is heavier than the sum of the three masses into which it decays, therefore conservation of energy requires that the mass loss shows up as the three final particles' energies of motion. Now, suppose instead of neutron beta decay, we consider the *forbidden* beta decay of a proton into a neutron positron and neutrino:

$$\text{Proton beta decay:} \quad \mathrm{p} \rightarrow \mathrm{n} + \mathrm{e}^{+} + \nu.$$

This process is energetically forbidden for a free proton because the proton is *lighter* than the neutron into which it decays. Putting it another way, the total energy of motion of the three final particles resulting from the proton decay would need to be negative to compensate for their greater mass compared to the original proton. However, if one of the three emitted particles (the neutrino) were a tachyon, the forbidden process could take place, because a tachyon's energy of motion can be negative for some observers. The LT helps us to explain why.

If Figure 2.6 looks familiar, that is because we had almost the same diagram in Figure 2.5, when we explained how the direction of a tachyon in time can become reversed when we switch observers. The left panel of Figure 2.6 shows a spacetime diagram for a "stationary" observer. Such observers are said to be "stationary" because they travel along with a rapidly moving proton, and therefore view it at rest. In the right panel we see how things look to a rapidly "moving" observer who watches the proton zip past. According to the stationary observer the proton decay yields a negative energy neutrino

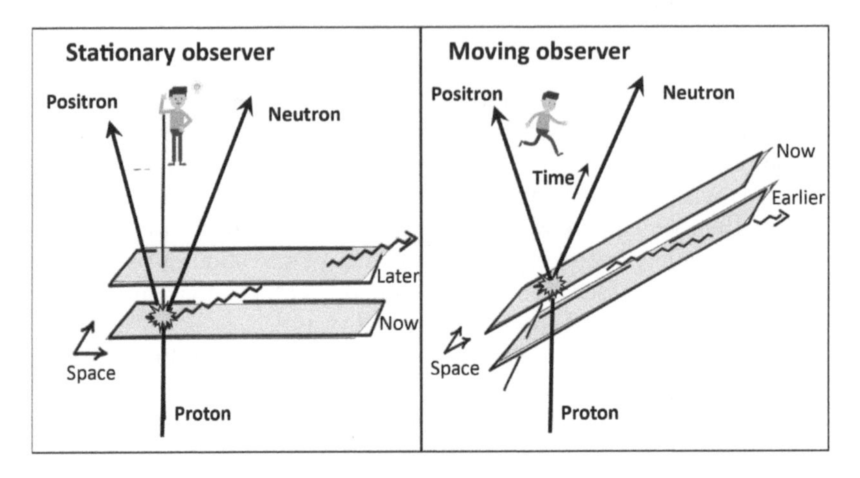

FIGURE 2.6
Protons can decay for the "stationary" observer (who sees proton at rest) if neutrinos are tachyons, and the proton energy exceeds some threshold.

(the squiggly line) which is *emitted* along with a neutron and positron (an electron's antiparticle). The moving observer, however, with the slanted space and time coordinates, sees the neutrino's time direction reversed. For the moving observer, the neutrino/tachyon is instead seen as an *incoming* positive energy particle that collides with the proton and creates a neutron and positron. This switch from an emitted negative energy neutrino to an absorbed positive energy (anti)neutrino is an example of the reinterpretation principle mentioned earlier. However, it is important to note that for the time reversal to occur, the proton needs to be moving at a very high speed. This requirement means that we would never see a free stationary proton undergo beta decay, because that would indeed violate energy conservation. *In summary, if we seek evidence that neutrinos are tachyons, just look to see if protons undergo beta decay when they have an energy above some very large value.* Given that protons in the highest energy particle accelerators show no evidence of decaying, if the process really is happening, we must look for it at still higher energies. The very highest energy protons ever observed are seen in the so-called cosmic rays produced by nature's own "accelerator."

Messengers from Space

Cosmic rays are particles incident on the Earth from space. Although they have been found to be a mixture of various nuclei, on average 87% of them are the simplest nucleus of all, namely protons. These cosmic ray particles

FIGURE 2.7
Drawing of cosmic ray showers from particles striking the top of the Earth's atmosphere.

Source: © INFN (f. Cuicchio/S lab)

often create "showers" of numerous secondary particles after colliding with air molecules at the top of the atmosphere – see Figure 2.7. Those showers might include thousands, millions or even billions of secondary particles, depending on the energy of the primary cosmic ray. Some of those shower particles can be observed by detectors placed on the ground, and they tell us about the nature, direction and energy of the primary cosmic ray particle that created the shower. Note that when such a shower is created the primary particle itself is not observed and identified. As indicated in Figure 2.8, the number of primary cosmic rays reaching Earth's atmosphere per second drops rapidly with increasing energy.

We see in Figure 2.8 that the log of the number of cosmic rays plotted versus the log of their energy is nearly a perfect straight line over ten decades of energy. The only departure from a perfect straight line in the figure is the very slight bend or abrupt steepening of the straight line that occurs right near the center of the graph. This feature, known as the *knee* of the spectrum, occurs at an energy of about 3×10^{15} eV, or 3 million GeV. For perspective, one GeV is about the rest energy of a proton. As Figure 2.8 shows, only one cosmic ray per *year* is found to strike each square meter of horizontal surface at the energy of the knee. Clearly very large area detectors are needed to observe appreciable numbers of cosmic rays at this energy or higher.

It may seem odd making a big fuss over the very slight change in slope of the spectrum occurring at the knee, but as can be seen in the figure, this feature is the most prominent one in the whole spectrum. What is the meaning of the knee? It is not unreasonable to assume that the spectrum of cosmic rays *at their source* was described by a perfect straight line, that is no knee. In that

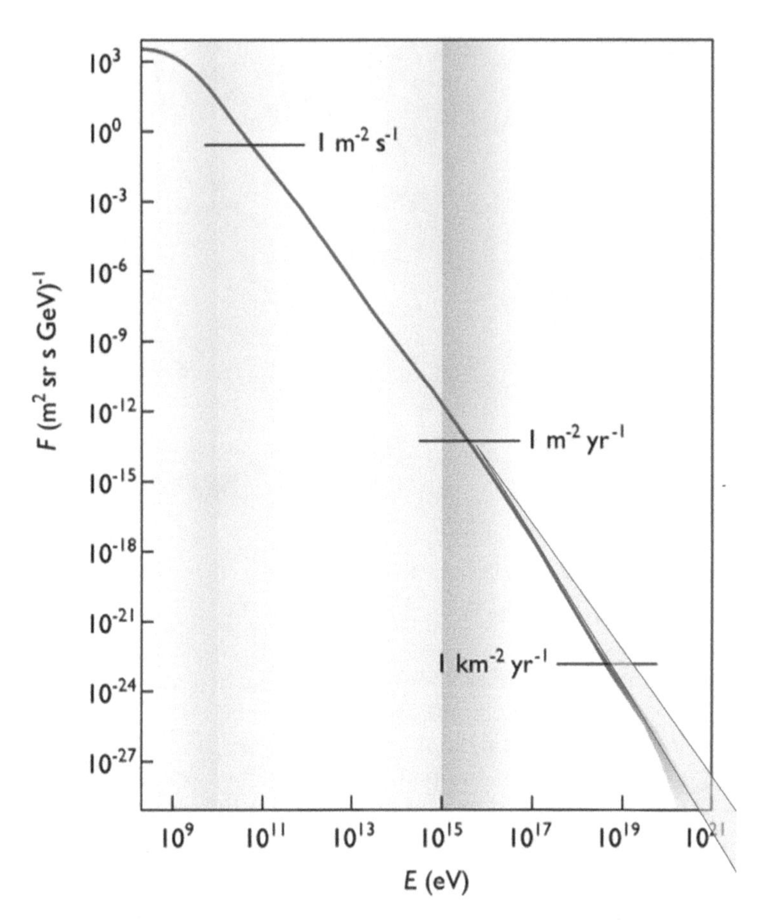

FIGURE 2.8
Cosmic ray flux versus energy in eV.

Source: Sven Lafebre, yellow triangle added by author. CC-by-SA 3.0 unported license

case, the knee could be created if beginning at an energy of 3 million GeV some process was removing cosmic ray protons from the spectrum in ever increasing numbers, with the shaded yellow area in Figure 2.8 representing the growing extent of the depletion with increasing energy. This interpretation of the knee would follow naturally if the neutrino were a tachyon. That's because protons would begin to beta decay above some threshold energy, and the fraction depleted would steadily increase at higher energies. The value of the threshold energy depends on the value of the tachyon-neutrino's mass. If for example, their mass squared were around $m^2 \approx -0.25$ eV^2 the threshold for proton decay would be 3 million GeV, namely right at the knee. Of course, this explanation of the knee is a bit circular, because we have no basis for assuming that the original spectrum of cosmic rays at their source should be a perfect straight line, on such a log-log plot.

Not surprisingly, the preceding explanation of the knee of the spectrum advanced by me in a 1999 paper has been accepted by approximately *zero* cosmic ray researchers. The usual explanation of the knee is that it is the transition from cosmic rays of a galactic origin to those having an extragalactic origin, whose spectrum has a slightly steeper slope. However, with this standard explanation, no agreement exists as to what mechanism accounts for the difference in slope above and below the knee, and at least seven possibilities have been suggested. Providing some additional support to my explanation of the knee, cosmic ray researchers Anatoly Erlykin and Arnold Wolfendale have claimed that the sharpness of the knee makes the standard explanation implausible. One possible way to settle the matter of the true cause of the knee might be to determine the directions of individual cosmic rays above and below the knee to see whether they originate from sources within or outside of our galaxy. However, as we shall see, finding the location in the sky from which cosmic rays originate turns out to be virtually impossible in most cases, unless they are neutral particles.

Locating the Sources of the Cosmic Rays

At low energies cosmic rays are produced in processes at or near the Sun, but at higher energies no one really knows their source for sure. The main problem in identifying specific cosmic ray sources more distant than the sun is that over very long path lengths charged cosmic rays cannot be traced back to their sources, but they have a direction that is randomized by the spatially varying magnetic field of the galaxy. Stable neutral particles are mostly unaffected by the galaxy's magnetic field and would point back to their sources, so neutral cosmic ray particles are the key to locating their sources, but as we shall see, definitive conclusions remain elusive.

Aside from neutrinos and photons, the only other relatively long-lived neutral particle known is the neutron, which has a lifetime of around 15 minutes. This lifetime, however, is not long enough for them to reach us from very distant sources. For example, if a neutron were traveling at the speed of light, you might think it would travel a mere 15 light-minutes or 0.00003 light-years before decaying (which is about the diameter of Earth's orbit). But there is a way that neutrons could travel much further distances if they had very high energy. In relativity, the phenomenon known as time dilation occurs, which says time slows down for moving "clocks." These include not only actual clocks, but also any process that occurs in a specific time interval, such as decaying neutrons, with their 15-minute half-life.

In relativity the factor by which time slows down for a moving neutron is given by the ratio of its energy E to its rest mass. For a neutron (or proton), as already noted, the rest mass is close to a GeV, which is a billion eV. Therefore, at the knee of the spectrum (3 million GeV) the lifetime of neutrons would be extended 3 million-fold, so that during one half-life they could travel about 100 light-years. Such a distance is much further than the nearest stars, however, no possible cosmic ray sources are found within that distance. As a result, even with time dilation, we would not expect any neutrons able to reach Earth at the energy of the knee. On the other hand, if neutrinos are tachyons, we shall see there may be a conceivable way for neutrons to reach us from very distant sources.

The Mysterious Cygnus X-3

Cygnus X-3 is a binary star, that is, two stars orbiting each other, and is one of the strongest X-ray and radio sources in the sky. In fact, during its strong radio flares it is the brightest radio source in the whole Milky Way galaxy. Cygnus X-3 is also notable because decades ago, there were numerous reports of cosmic rays coming from its direction with energies up to the knee region, and this directionality would imply that the cosmic rays from this source were neutral particles. Not only were the particles coming from the direction of Cygnus X-3, but they also seemed to be arriving in synch with its rotation, as if a giant rotating searchlight beam were sweeping past the Earth once each 4.8-hour Cygnus X-3 rotation.

The mystery here is that the only three neutral particles that could reach us from Cygnus X-3 would seem to be neutrinos, gamma rays, and neutrons, but none of them seemed possible. Neutrinos could be ruled out, because these weakly interacting particles would not produce the large number of events seen in the detectors. Gamma rays could be ruled out, because the showers they would create in the atmosphere would contain many more electrons than were observed. Neutrons also seemed to be ruled out, because as previously noted, given the distance to Cygnus X-3 (about 25,000 light-years), neutrons with an energy near the knee would decay long before reaching Earth. Still, there was some circumstantial evidence favoring neutrons: namely the deep underground muons found to accompany the showers, which were a sign that the primary cosmic ray was a neutron not a gamma ray.

The Proton–Neutron Decay Chain

Here we shall see how neutrons might survive the trip from Cygnus X-3, but the idea depends on neutrinos being tachyons. Recall that proton decay would be possible for energies above the knee if neutrinos are tachyons –

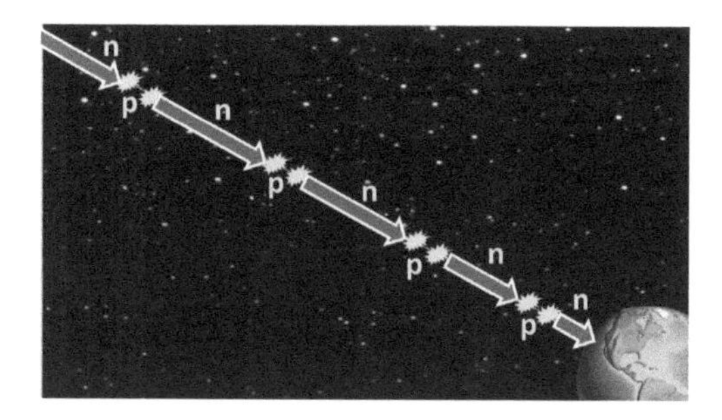

FIGURE 2.9
A proton–neutron decay chain, which could occur when incident cosmic ray protons have energies above the knee if neutrinos are tachyons

which was how we explained the origin of the knee. In this case, it is possible to have a hypothetical "proton–neutron decay chain," n→p→n→p…, (see Figure 2.9). Note that in writing this process the way we have, we are ignoring the electrons, positrons, and neutrinos produced at each step of the chain during beta decay. Such a decay chain would be allowed provided a proton's energy is above the threshold for p→n decay (assumed to be at the energy of the knee). At each step of the hypothetical decay chain, the decaying neutron or proton loses a bit of energy, because the other two particles in the decay carry some away, but the n or p would mostly maintain its original direction it had before decay. Moreover, during the decay chain the cosmic ray spends most of its time as a neutron rather than a proton because the p→n decay would be much quicker than n→p. Thus, for cosmic ray protons starting out from Cygnus X-3 having initial energies above the knee we would expect them to behave mostly like neutral particles (that is, neutrons not protons), and hence they would roughly point back to their source, and they would reach Earth with energies close to that of the knee.

In summary, if the neutrino is a tachyon, then based on the hypothetical decay chain, we might expect an enhanced number of cosmic rays from the direction of Cygnus X-3 having an energy just below that of the knee. In a second 1999 paper I reported finding exactly such an enhancement in the number of cosmic rays from the direction of Cygnus X-3 using previously published data. This second paper was greeted with even more skepticism than my earlier 1999 paper, which had claimed the knee of the spectrum was due to neutrinos being tachyons. Aside from skepticism about tachyons, there were additional questions raised about whether Cygnus X-3 had in fact ever really been a source of cosmic rays – despite the claims to that effect by a *dozen* different cosmic ray groups. These doubts arose when several new more sensitive detectors failed to see any excess cosmic rays from the direction of Cygnus X-3. Most researchers were understandably skeptical that Cygnus X-3 could be a "shy" source that conveniently turned-off just when more sensitive detectors became available!

An Unsettled Mystery

The consensus among nearly all cosmic ray physicists today is that despite the dozen reports of cosmic rays from Cygnus X-3 from several decades back, those earlier sightings were just a mirage. The most sensitive of the newer detectors, known as CASA-MIA, had accumulated over one hundred times as many events as earlier experiments, and it reported no hint of any cosmic ray excess above background in the direction of Cygnus X-3. Moreover, Jim Cronin, the head of the CASA-MIA collaboration was a widely respected physicist at the University of Chicago and a Nobel Laureate, so it is unlikely their detector had missed something. After their data was published I contacted Cronin, and he very kindly allowed me to visit his group and examine the original data, so I could see for myself that there was indeed no sign of any signal associated with Cygnus X-3. After spending a freezing winter weekend in Chicago looking at their data, I finally realized what the explanation of the missing signal might be. Due to the geometry of the CASA-MIA detector much less than 0.1% of its data were taken at or above the energy of the knee of the spectrum. If my theory of a neutron-proton decay chain was right, and the cosmic rays from Cygnus X-3 therefore mostly had an energy just below the knee, obviously CASA-MIA with so little data in that small energy region could not possibly show any signal from that source. Their negative result therefore would not contradict previous positive ones that were sensitive to that energy region.

Thus, to summarize the previous discussion, despite the conventional wisdom of cosmic ray researchers, I believe the issue of whether Cygnus X-3 was ever a genuine source remains unsettled. The key to the puzzle is to recognize that any excess number of cosmic rays from the direction of Cygnus X-3 would be found only if one selected *both* (a) a narrow time interval in synch with the rotating source and (b) a narrow energy interval in just below the knee of the spectrum. Unless one makes these two selections on the data simultaneously, *and* enough data remain to make a proper test, one would not expect to see excess numbers of cosmic rays from the direction of Cygnus X-3. The newer experiments reporting no excess of events from the direction of Cygnus X-3 either lacked enough data in the knee region, or they did not make the above two data selections, so their negative results did not really contradict earlier positive ones.

A Failed Collaboration: *Glupyy Amerikanets!!*

Following my work on Cygnus X-3, I reached out to people in several cosmic ray groups to see if their data might show evidence of cosmic ray sources. I was hoping to find evidence for an enhanced number of cosmic rays with energies near the knee from any specific directions in space, not only that of Cygnus X-3. One person I contacted was Mikhail Zotov, a physicist at Moscow State University, and a member of the Tunka Cosmic Ray Collaboration. Tunka, a large cosmic ray detector located near Lake Baikal in Siberia, was well-suited to look for evidence of cosmic ray sources at the energy of the knee. In fact, Mikhail had previously been looking for just that possibility, and we began collaborating. After we spent over a year analyzing the Tunka data, I became convinced that there was indeed evidence for cosmic ray sources in the data, based on finding some small regions of the sky showing enhancements in the number of cosmic rays near the knee's energy. Mikhail, however, had his doubts, and after I wrote up a draft, he declined either to be a coauthor on the paper or to offer specific suggested changes on it. On the other hand, he also did not express any objection to my submitting the paper on my own.

This paper I submitted was rejected, in part because its conclusions were not supported by any member of the Tunka Collaboration. I suspect that Mikhail's decision to decline to be a coauthor may in part have been due to his discomfort with tying observations that coupled the idea of cosmic ray sources to my tachyon neutrino hypothesis. Making a claim of the existence of specific cosmic ray sources would have, by itself, been enough of a challenge to the conventional wisdom, which says none have yet been found. However, coupling such a claim to some craziness about tachyons was perhaps a step too far for Mikhail. Confirming my suspicion about the reasons for the breakdown of our collaboration, Mikhail later wrote a similar paper analyzing the Tunka data which identified many possible cosmic ray sources in the knee region, without any mention of tachyon neutrinos as the possible explanation.

In retrospect, I probably should not have been too surprised at the collapse of our collaboration because Mikhail had earlier confided to me that his previous collaboration with an American researcher did not work out. The whole episode may also have been due to an unfortunate misunderstanding on my part, and my lack of appreciation for the hierarchical nature of cosmic ray groups, especially Russian ones. When Mikhail and I began our collaboration, I had been under the naïve impression that it was with the blessing of the head of the Tunka group, and that we would be free to publish anything interesting that we found, but such a decision in fact was probably never ours to make, and certainly not for me to make on my own after Mikhail withdrew. By the way, the title of this section, *"glupyy amerikanets,"* is the transliteration of "Stupid American," based on my self-assessment, not anything Mikhail ever said.

My unpublished paper analyzing the Tunka data can still be found on the e-print archive (along with Mikhail's own version), and perhaps some other cosmic ray group might be interested in examining their own data using the same method that the paper describes. In fact, one need not be a member of any group to do such an analysis, because nowadays some cosmic ray collaborations store all their data online, and they allow open access to it. This open access is granted after the group has already published their results, and it allows others to look for features in the data that the group may not have considered.

Some readers might be wondering why I have not made any further efforts to check out other cosmic ray data on the web myself. The answer is that I have found other approaches to the problem of tachyonic neutrinos much more promising than cosmic rays. Additionally, looking back now two decades later, I must admit to having severe misgivings with respect to my previous cosmic ray work. Although this work may have been a useful first step on the road to finding evidence that some neutrinos are tachyons, ultimately it was unsatisfying. This negative assessment is due in part to the still unresolved issue about whether Cygnus X-3 or any other objects were ever cosmic ray sources at the energy of the knee. In addition, there is the matter of the many alternative explanations of the knee of the cosmic ray spectrum, all of which are less exotic than neutrinos being tachyons. Finally, my concerns relate to the very nature of the field of cosmic ray physics itself.

The Field of Cosmic Ray Physics: A Work in Progress

The field of cosmic rays is over a century old, arising in 1912 when Victor Hess carefully measured ionizing radiation levels from a gondola attached to a hot air balloon as it rose above the Earth. Hess correctly concluded he had discovered some form of "penetrating radiation" coming from space, a discovery that earned him a share of the Nobel Prize in physics some 24 years later. Subsequent researchers have continued to ascend hot-air balloons, as well as climbed mountains, launched satellites, and traveled to the far corners of the Earth and beyond to better understand these fast-moving messengers from space. Still, after more than a century of cosmic ray studies, most of the main questions about the cosmic rays are still unresolved. As one cosmic ray theorist, Alvaro de Rujula, has noted, "the number of correct predictions in the field is impressively small."[10]

Thus, we do not yet know where in the sky cosmic ray sources are located, exactly what those sources are, how they accelerate the particles to their observed energies, what the cosmic ray composition is, and how they propagate to reach us. The mystery is particularly acute for cosmic rays having the very highest energies, above 5×10^{19} eV, which never should have reached Earth in the first place according to theory. Above that energy, known as the GZK cutoff, cosmic rays should be increasingly blocked by the cosmic background radiation filling all space, and yet they somehow come. Not only do they reach us, but their composition is completely unknown; they could be neutrons, protons, heavy nuclei, or even some unknown particle.

One cosmologist, Luis Gonzalez-Mestres, has even suggested that these very highest energy particles above the GZK cutoff are superluminal or FTL.

It is striking to compare the limited ability to answer fundamental questions in the field of cosmic rays with that in particle physics – a field that grew out of cosmic ray physics. Particle physics has its well-developed standard model (see Chapter 4), which has made many verified predictions regarding the existence and properties of new particles. In contrast, cosmic ray physics, a far older field, has no fundamental theories, and mostly consists of a collection of conventional wisdom. There are many reasons for this difference between the two fields, reasons which are attributable to accidents of nature, rather than any lack of ingenuity of cosmic ray researchers. These include the unfortunate facts that charged particles do not point back to their distant sources, and their detection is necessarily indirect, being only through the secondary particles in the showers they produce. In addition, it is difficult not only to discern the primary particle's identity, but even whether the particle is charged or neutral.

Another problem with solving cosmic ray puzzles is that one must bring in ideas from many other fields, including astronomy, cosmology, astrophysics, geophysics, nuclear physics, and elementary particle physics. As a result, if some anomaly is found in the data just where to look for an explanation is unclear. Thus, for example, as we have seen, the knee of the spectrum has too many explanations, not too few, so we really cannot be certain just what is the basis of this most significant feature of the cosmic ray spectrum. In view of the many conventional explanations for the knee, an oddball model involving tachyonic neutrinos would not likely receive a warm reception by cosmic ray physicists.

INTERMEZZO

In the remainder of this chapter, we temporarily leave the topic of seeking evidence of tachyons, or more specifically that some neutrinos are tachyons. Here we consider some "fun topics," particularly the paradoxes that tachyons and time travel would create, and how these matters are treated by philosophers, science fiction writers, and physicists. We also consider the wormhole, a hypothetical entity that physicists think is the most likely vehicle for both time travel and FTL space travel.

Although time travel to the past raises various paradoxes, this is not the case for time travel to the future, based on the effect known as time dilation, discussed earlier. Thus, for example, here is a simple way to travel 1000 years into the future – at least in principle. Just catch a ride on a spaceship that traveled to a star 500 light years away at an average speed of 99.99995% the speed of light, and then return to Earth at the same average speed. People on Earth would find your outbound trip lasted 500 years, the return trip 500 years, for a total of 1000 years, but for you aboard the ship it would last just 1 year. It is time travel to the past where the paradoxes arise, which are similar to those arising when sending messages to the past.

Contacting Your Earlier Self

A hypothetical tachyon-based device for communicating with the past has been called an *antitelephone,* a term coined by Gregory Benford.[11] See Figure 2.10 for an artist's conception of such a device. Let us see how you might contact your past self, using tachyons.

Suppose Alice at age 30 wants to send her earlier self an important message, possibly concerning an ill-chosen romantic partner. She first sends the message encoded as a tachyon signal to Bob on a distant spaceship, asking him to transmit the original message back to her as soon as he receives it. Bob's spaceship is 10 light years from Earth and moving away from it. Amazingly, Bob's relayed tachyon signal back to Alice could reach her before she sent the original one to him! An analysis of this scenario using the mathematics of the LT shows that Alice could receive the relayed signal from Bob when she was as young as 20, that is, 10 years before she sent the original signal.

A logical paradox then arises, because she might decide ten years later not to send Bob the original version of the message when she reaches age 30. In that case, what was the origin of her earlier received message? One resolution of the paradox would be that free will is just an illusion, and she simply cannot make such a decision. Einstein himself was a determinist who held that: "Human beings, in their thinking, feeling and acting are not free agents but are as causally bound as the stars in their motion."[12] Recent psychological research may support Einstein's apparent belief in the illusory nature of free will. Experiments have shown that the brain convinces itself that it made a free choice from the available options after the decision has been made. In other words, once we see what we have done, we believe that our decision

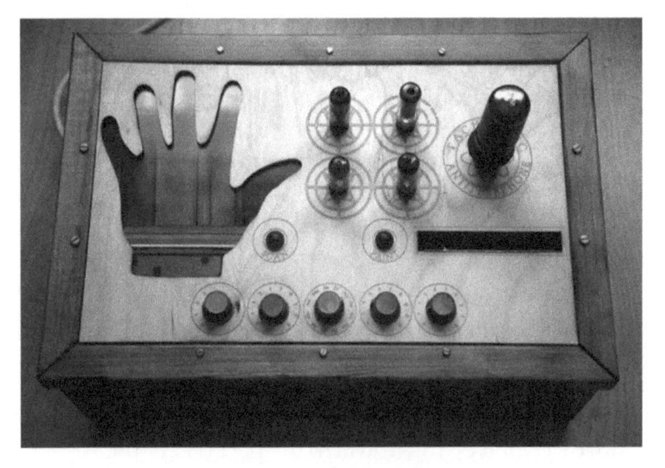

FIGURE 2.10
Conception of a tachyon antitelephone provided by Dublin artist Sinead McDonald.

was based on our (imagined) free will. Of course, Einstein and most other people who accept the determinist view live their lives as though they had free will, even though it may be a persistent illusion.

A "More Feasible" Way to Contact the Past

The idea of Alice contacting her 10-year younger self with the aid of Bob in a spaceship 10 light years away clearly is unfeasible, so here we consider a simpler way, depicted in Figure 2.11. The figure shows 13 satellites all in the same orbit, and they rotate about Earth in a clockwise sense.

The message you wish to send to your past self is first transmitted from Earth using tachyons to one of the satellites. That message is then instantly relayed to an adjacent satellite using tachyons, which again instantly relays it to the next one and so forth, until the message returns to the first one which relays it back to Earth. Note that the rotation direction of the relayed messages is also clockwise – the same direction the satellites are moving. According to relativity, the return message can arrive back at the original satellite earlier than the original message was sent by an amount that depends on the tachyon and satellite speeds. The math here is very similar to the earlier example of chasing a tachyon. Thus, for example, if we used tachyons having a speed 437 times that of light, the return tachyon signal will be in the past provided the speed of the satellites in their orbits exceeds 1/437th

FIGURE 2.11
Scheme for sending a message to the past using an orbiting ring of 13 satellites.

the speed of light. To send a message to the distant past, one would need to have the tachyon message make many circuits. For example, suppose the circumference of the satellite ring was 300,000 km, so it took a light signal one second to complete the circuit around it. If you wanted to send a signal back to your 10-year younger self, the minimum possible circuits around the loop would be about 310 million, which is the number of seconds in 10 years.

The actual feasibility of this "more feasible" scheme is unknown and would depend on many factors, most importantly whether tachyons exist! Moreover, even if tachyons do exist, it is quite possible that all schemes to use them to contact the past are completely impractical. Thus, in our previous example using tachyons having a large multiple of light speed, if the tachyons are neutrinos, they would necessarily have extremely low energy, because of the way energy depends on speed for tachyons. As a result, such very low energy neutrinos would be entirely undetectable with any known technology. On the other hand, we do need to remember that Wolfgang Pauli, the father of the neutrino, thought he had postulated an unobservable particle, so who knows if neutrinos having very low energy might someday be detectable.

Fictional Examples

There are many examples of paradoxes involving tachyons in science fiction stories. For example, scientists in Benford's "Timescape," uses tachyon messaging to avert an environmental disaster when scientists attempt to alert colleagues living in an earlier era of a dire situation. Here is a succinct description of the sort of logical paradox tachyon antitelephones can lead to, as described by Benford.[12] Suppose you agreed to send your friend a message at 5 PM only if you do not receive one from her at 4 PM. Let's say you do receive a message from her at 4 PM. In this case, you do not send her any message at 5 PM, then the exchange of messages will take place if and only if it does not take place.

This paradox (and the previous one with Alice sending a message to her earlier self) is the tachyon version of the classic one about going back in time and killing your grandfather. Were you to accomplish the deed, you would prevent your own existence, making it impossible for you to have traveled back in time in the first place. Some physicists have concluded that neither time travel nor FTL tachyons can exist because of such paradoxes. An alternate view has been proposed by physicist Igor Dmitriyevich Novikov with his self-consistency principle. According to this principle, time travel or communication with the past might be possible, but only self-consistent situations would be permitted. Thus, if you went back in time to kill your grandfather you would not be able to do so, since that would be incompatible with your existence. Essentially, you would always be frustrated in any attempt to kill the younger version of the old man. This has been called the "banana peel" solution to the paradox. No matter what you do something happens, like slipping on a banana peel that just happens to be conveniently placed so that the paradoxical event fails to occur. Alternatively, you would always simply change your mind about doing the dastardly deed. Finally, if you did succeed in killing a person you thought was your future grandpa, it would turn out that he was not your real grandfather.

You might think that it would be futile for scientists in the Timescape story to advise their past colleagues on how to avert an environmental disaster, given the impossibility of changing the past. However, in the story the future scientists, aware of the need to avoid a paradox, send garbled messages, giving the past researchers just enough information to start efforts on solving the pending ecological crisis, but not so much that the crisis will be entirely solved. If the crisis had been solved, a signal to the past would have been unnecessary, and it never would have been sent. An alternative solution to the grandfather paradox is that the requirement for self-consistency is wrong, and we really *can* change the past. In this case, we might enter a parallel universe or a different timeline if a message from the future were to change the past.

In time travel stories when time travelers visit previous or future times during their own lifetime, they sometimes encounter multiple copies of themselves. The ultimate example of this idea is Robert Heinlein's story "All You Zombies," which was made into the movie "Predestination." The paradoxical story starts when a baby girl "Jane" is mysteriously dropped off at an orphanage in Cleveland in 1945. The story involves multiple time travel journeys to the past involving what seems to be four major characters, but as the story unfolds, it turns out that all the four major characters are the same person, at different stages of their life, and even having different genders. Jane is revealed in the story to have been born intersex, and after sexual reassignment surgery he is taken back in time, and impregnates his younger self, giving birth to another version of Jane. If we draw Jane's bizarre family tree, all the branches curl back on themselves, and we are led to the astonishing conclusion that Jane is her own mother and father! Yet, the story is an example of one that is paradoxical and yet perfectly consistent.

Perhaps the most elegant of the possible solutions to the various paradoxes that time travel and backwards time signaling create has been suggested by science fiction writer Larry Niven. According to "Niven's Law" of time travel

> If physical laws governing our universe permit the possibility of backwards-time signaling or time travel which can change the past, then no devices for doing either one will be invented.[13]

In other words, a message sent to the past alters the entire future history, including the sending of the original message *and* its content. The altered message will change the past in a different way, and so on, until some "equilibrium" is reached – the simplest being where no message at all is sent. Backward time signaling thus erases itself.

Wormhole Time Machines

If time travel is possible, the most discussed mechanism has been the wormhole. Wormholes are a theoretical solution to the equations of *general relativity*, which is Einstein's theory of gravity that was published ten years after his 1905 special relativity theory. According to Einstein, gravity is not a force but rather a distortion of spacetime caused by the presence of matter – the greater the density of matter, the greater the distortion nearby. For a simplified notion of a wormhole, visualize space as a folded over two-dimensional surface – see Figure 2.12. In this case, a wormhole might consist of two holes or mouths in that surface, with a connecting tube or tunnel. The two wormhole mouths could be separated either spatially or temporally by arbitrarily

FIGURE 2.12
A model of a wormhole connecting two regions of flat space-time.

Source: Image by Shutterstock

long distances or times. In other words, the trip through the tunnel could be a shortcut in spacetime connecting two remote times, remote places, or possibly even two universes.

In 1992, Stephen Hawking wrote a paper about his "chronology protection conjecture," in which he acknowledged that wormholes could exist, but he doubted they would be stable, so that history would therefore be forever "protected" from time travelers. However, Hawking apparently changed his mind, based on his posthumously published 2018 book *Brief Answers to the Big Questions*.[14] In the book Hawking expressed the view that FTL space travel and travel back in time cannot be ruled out. Moreover, research in 2021 by Juan Maldacena and Alexey Milekhin of the Institute for Advanced Study in Princeton has shown that humanly traversable wormholes are consistent with the laws of physics as we presently know them. In addition, while it had earlier been thought that stable wormholes would require exotic matter having negative mass, recent research has shown this may not be necessary.

Where Are the Time Travelers?

Stephen Hawking (see Figure 2.13) at one time expressed his puzzlement about the absence of time-travelers from the future if time travel really were possible. In fact, he considered their absence one of the strongest arguments against the possibility of time travel. Hawking, who was a theorist, once conducted an actual "experiment" by holding a party for time travelers just to see if any showed up. What time traveler from the future could resist such an opportunity to meet with the eminent Hawking? To ensure that any attendees were true time travelers he sent the invitations out *after* the party was held, so only they would know about it. Not surprisingly, there were no attendees to his party, which certainly shows that Hawking had a great sense of humor, but it says nothing, of course, about the possibility of time travel. Any time travelers who might have wanted to attend knew very well they could not do so, since there was no record of anyone but Hawking present, so they could not have blended in with other attendees.

Clearly, Hawking's party for time travelers was simply a joke. On the other hand, according to The Stephen Hawking Foundation, time travelers were also invited to the memorial ceremony following Hawking's demise. Given that 1000 tickets were made available to the public, one cannot rule out the remote possibility that some time travelers attended that event. If time travel really is possible, the simplest explanation as to why no travelers have been seen is that it is only possible to go back to past times after the time machine has been invented, which has not yet happened. On the other hand, it will probably not surprise you to learn that there are many examples of people claiming to be time travelers. Many amusing (but not very persuasive)

FIGURE 2.13
Stephen Hawking in 1992 at a London press conference.

Source: Shutterstock image

accounts of such people can be found on the web. Presumably, if there were an occasional real time traveler among them who wanted to convince you they were the real deal, this would not be terribly difficult if they just made a very large number of predictions, every one of which turned out to be true.

Most of what has been published on wormholes is in the theory realm, especially how they might form, and whether they could be stable enough for passage through them. Currently, it is believed that very tiny wormholes do exist, although they have never been observed. At the very smallest distance scales (the so-called Planck length, or about 10^{-35} m) it is believed that spacetime would consist of many small, ever-changing regions in which space and time are not definite, but they fluctuate in a foam-like manner. This "quantum foam" was first suggested by physicist John Wheeler, based on the Heisenberg Uncertainty Principle. Some foamy regions might be in the form of tiny natural wormholes. It is possible that such tiny wormholes in the very early universe could have been enlarged naturally by many trillions of times during the so-called inflation era in the very early universe. It is even conceivable that such naturally formed wormholes might now be big and stable enough for a human or a spaceship to pass through. However, this is all pure speculation. Barring such a natural formation, it might be possible in the future to capture a tiny wormhole, stabilize it, and enlarge it. One might even imagine that some super-advanced aliens have already figured out how to do this, and they have planted many wormholes around the galaxy, much like metro stops.

Searching for Wormholes

Even though wormholes are theoretical solutions of general relativity, we have no idea if they really exist, but this has not stopped astronomers from looking for them. For example, four Argentinian astronomers looked for the very peculiar type of gravitational lensing of light from distant background sources, which might result from a foreground wormhole. An example of gravitational lensing (but almost certainly *not* by a wormhole) is illustrated in Figure 2.14.

The figure shows a near-complete ring, which is the distorted image of a background galaxy by the gravity of a large red foreground galaxy. We can get an idea of the kind of gravitational lensing that a wormhole might create through a simulation created by astrophysicist Corvin Zahn, which is illustrated in Figure 2.15. This simulation shows what would be seen when

FIGURE 2.14
An "Einstein Ring" – an example of gravitational lensing of a large red galaxy LRG 3-757.

Source: NASA/Hubble image, pub. Domain

FIGURE 2.15
Imaginary wormhole.

Source: Author: Corvin Zahn, Hildesheim University, Germany

looking into one mouth of a wormhole located at the physical institutes of Tübingen university. It has been assumed that the other mouth emerges at the sand dunes near Boulogne sur Mer in the north of France, which you can see in the central part of the image. Passing through the imaginary wormhole you could travel from one location to the other in practically no time at all.

Regarding naturally occurring wormholes, some astronomers have suggested that the supermassive black holes at the center of every normal galaxy might really be wormholes created in the early universe. Still others have claimed that wormholes might reside not just at the galactic center but in many parts of our galaxy including our own galactic "neighborhood."

A Wormhole in the Solar System?

A solar system wormhole was the premise of the sci-fi movie *Interstellar* in which advanced aliens put a wormhole near Saturn so that humans who had wrecked their home planet could use it to make the interstellar journey to a

FIGURE 2.16
Kip Thorne, co-recipient of the 2017 Nobel prize in physics.

Source: Photo by Ed Carreon

new home. Placing one entrance to the wormhole near Saturn may have been their way of saying "if you have made it this far, you have a good chance of surviving the trip through this wormhole we've put here for you." Even if wormholes do exist, voyages through them will likely remain in the realm of science fiction, according to astrophysicist Kip Thorne of the California Institute of Technology in Pasadena (see Figure 2.16). Thorne, who served as an adviser and executive producer on *Interstellar* has also written the book, *The Science of Interstellar*.[15] He believes that if traversable wormholes exist, then almost certainly they cannot occur naturally, but would be created by an advanced alien civilization. So, maybe finding one in the solar system is not completely crazy. In the *Interstellar* movie the advanced alien civilization that put the wormhole there turns out to be our descendants who visited their past to save us from extinction. Their visit of course was not entirely altruistic, since for them to exist we had to be saved!

Our ability to travel to Saturn will probably be decades in the future. So, even if we took the *Interstellar* plot seriously, it might be prudent to try to fix some very serious Earthly problems (including pandemics, climate change, giant solar flares and nuclear war), rather than counting on future humans to travel back in time to rescue us with a wormhole. Still, when I think the world has gone completely insane, the idea of escaping to another part of our universe or possibly a parallel universe through a wormhole has a certain attraction. On such occasions I can fantasize about the pretend wormhole rug in my own home – see Figure 2.17.

FIGURE 2.17
Wormhole rug in the author's home office.

Parallels between Wormholes and Tachyons

Tachyons and traversable wormholes, science fiction favorites, have many similarities. Both are hypothetical entities consistent with the equations of relativity (general relativity for wormholes and special relativity for tachyons). Many physicists are dubious whether either entity exists, but there is a wide spectrum of opinion on the matter. If they do exist, both entities might permit FTL travel – tachyons for the particles themselves and wormholes for humans using them to visit another star. Both tachyons and wormholes are also connected to backward time travel for messages or people. Finally, both came to the attention of many physicists through articles written in an educational journal, not a research journal, namely *The American Journal of Physics.* In 1988 that journal published the classic article "Wormholes in spacetime and their use for interstellar travel"[16] by Michael Morris and Kip Thorne (of *Interstellar* fame), which considered what would be necessary for a traversable wormhole. This article, cited an impressive 2106 times, established a strong link between the science fact and fiction of wormholes, just as the 1962 "Meta" Relativity article in the same journal did for tachyons.[17]

Summary

Our physical universe consists of a stupendous number of subatomic particles that make up all matter, including our bodies. The cosmic dance of these particles can be described using a four-dimensional spacetime diagram in relativity, where the past, present, and future of each particle is described by its worldline. Relativity describes how the observations of the same worldlines as seen by different observers relate to one another, based on the LT, which has a simple geometrical interpretation. Tachyons in a spacetime diagram have space-like worldlines that fall into the "elsewhere" region (outside the light cone), which allows them to travel backward in time, according to some moving observers. For example, for tachyons emitted from a transmitter or source and then later detected at a receiver, the time sequence can be reversed, which in effect switches the roles of transmitter and receiver. This switch, in principle, might allow you to send a message backward in time, and several ways are explored to send a message to your earlier self.

The switches in time direction that can occur for tachyons led me to some initial work in seeking evidence that neutrinos are tachyons, using published data for high energy cosmic rays. The idea is that when a switch occurs in a tachyons' direction in time, it is accompanied by a switch in the sign of its energy. Since tachyons can have a negative energy then certain forbidden processes can occur, such as the beta decay of a free proton into a neutron (plus an electron and a neutrino). But this process would only be allowed if a proton is observed moving at some sufficiently high energy or speed. This forbidden decay of very high-energy protons might conceivably explain the existence of the knee of the cosmic ray spectrum, but only if electron neutrinos are tachyons. Cosmic ray researchers, of course, have more conventional explanations for the knee, namely a shift from galactic to extragalactic sources that occurs at that energy.

In seeking further evidence in the cosmic rays for the notion that neutrinos were tachyons I came across the Cygnus X-3 mystery. During the 1970s and 1980s, there were numerous reports of neutral cosmic rays from the direction of the intense radio source Cygnus X-3, and the identity of these cosmic rays was very puzzling. There seemingly was no known neutral particle that could explain the observations. I showed, however, how these strange Cygnus X-3 observations could be explained based on a hypothetical n-p decay chain that would only be allowed if neutrinos were tachyons. However, this idea has not been accepted by most cosmic ray physicists, who have severe doubts that cosmic rays from Cygnus X-3 actually were ever really observed, despite nearly a dozen groups having reported them. Those doubts arose because more sensitive new detectors failed to see cosmic rays from Cygnus X-3. On the other hand, the new detectors had little sensitivity for cosmic rays with energies near the knee. Thus, contrary to the conventional wisdom, the Cygnus X-3 mystery may still be unresolved.

After an attempted (but failed) collaboration with a cosmic ray researcher connected with the Russian Tunka experiment to investigate cosmic ray sources, I abandoned my early tachyon hunting efforts based on cosmic ray data and moved on to more promising approaches. In the final part of this chapter, we considered the paradoxes that tachyons and time travel would create, and how these matters are treated in fiction. We also considered the hypothetical entity that physicists think is the most likely possible vehicle for both time travel and FTL space travel, namely the wormhole, and how such objects might be detected. The consensus view is that tiny wormholes probably exist, but it is uncertain if large ones do, and it is also unclear whether they could be made stable and traversable. Searches by astronomers for wormholes have so far been inconclusive. There is much debate in the physics community about whether time travel to the past using wormholes would be possible, and some luminaries like Stephen Hawking have changed their opinion on the matter of time travel from "impossible" to "maybe."

References

1. Quoted in a letter to the widow of Einstein's close friend, Michele Besso, in *Time's arrow: Albert Einstein's letters to Michele Besso,* https://www.christies.com/features/Einstein-letters-to-Michele-Besso-8422-1.aspx
2. Benford, Gregory. *Timescape,* New York, NY, Simon and Schuster, 1980.
3. As quoted in a conversation with Max Born about the development of the theory of relativity, by Carl Seelig, Albert Einstein: A Documentary Biography, Staples Press, London, 1956.
4. Snow, Charles Percy. *Variety of Men,* London, Penguin Books, pp. 100–101, 1966.
5. Minkowski, Hermann. "Space and Time," edited by H. A Lorentz, A. Einstein, H. Minkowski, and H. Weyl, *The Principle of Relativity: A Collection of Original Memoirs on the Special and General Theory of Relativity,* New York, Dover, p. 75, 1952.
6. Wells, Herbert George. *The Time Machine,* William Heinemann (UK), Henry Holt (US), 1895.
7. James, Ioan. *Remarkable Mathematicians: From Euler to Von Neuman,* Oxford, Mathematical Institute, p. 14, 2002.
8. Henderson, Bobby. The Gospel of the Flying Spaghetti Monster, New York, NY, Villard Books, 2006.
9. Lucas, John. *The Future: An Essay on God, Temporality and Truth,* Oxford, Basil Blackwell, p. 8, 1989.
10. De Rujula, Alvaro. The Cosmic-Ray Spectra: News on their Knees, Physics Letters B, 790, 444–452, 2019.
11. Benford, G. A., D. L. Book, and W. A. Newcomb. The Tachyonic Antitelephone, *Physical Review. D,* 2, 263, 1970.
12. Goldman, Robert. *Einstein's God: Albert Einstein's Quest as a Scientist and as a Jew to Replace a Forsaken God,* Lanham, MD, Jason Aronson publisher, 1997.

13. Nivens' law as described in Analog Magazine, January 29, 2002.
14. Hawking, Stephen. *Brief Answers to the Big Questions*, London, Hodder and Stoughton Publishers, 2018.
15. Thorne, Kip. *The Science of Interstellar*, New York, NY, W. W. Norton, 2014.
16. Morris, M., and K. Thorne. *Wormholes in Spacetime and Their Use for Interstellar Travel, American Journal of Physics*, 56, 395, 1988.
17. Bilaniuk, O. M. P., V. K. Deshpande, and E. C. G. Sudarshan. "Meta" Relativity, *American Journal of Physics*, 30, 718–723, 1962.

3

Supernova SN 1987A and Its Three Unicorns

The nitrogen in our DNA, the calcium in our teeth, the iron in our blood, the carbon in our apple pies were made in the interiors of collapsing stars. We are made of star stuff.[1]

Carl Sagan, Astronomer

Supernovae

Billions of years ago, the atoms of our bodies were cooked up in a massive star that underwent a massive explosion and spewed the nuclei of those atoms into space. We are literally "star stuff" in astronomer Carl Sagan's phrase. Stellar explosions known as supernovae occur in our galaxy only two or three times per century, with perhaps 90% of them obscured from view by interstellar dust. In this chapter we shall meet three "unicorns" present in the data from a supernova observed over three decades ago. Be aware that the word *unicorn* can refer either to "something unusual, rare, or unique," or alternately to a mythical creature, which we shall see is a particularly appropriate ambiguity.

Our tale of the three unicorns begins about 168,000 years ago when an unremarkable star known as Sanduleak-69 202 became a supernova in a satellite galaxy on the outskirts of the Milky Way. Earthlings first learned of this event when the light from the massive explosion, having traveled about 168,000 light-years through space, finally reached us on February 24, 1987. Because it was the first supernova observed that year, it received the designation "A," and so it is known as SN 1987A." Amazingly, the next one, SN 1987B, was observed the very next night. However, it occurred in a different galaxy, and was 10,000 times fainter, so little attention has been paid to it. The previous recorded supernova in our galaxy prior to SN 1987A occurred in 1604 AD, and it was so bright that it could be seen in broad daylight for more than three weeks. Since the invention of the telescope in 1608, many supernovae have been observed in our own and other galaxies, but none of them was visible to the unaided eye until SN 1987A. A particularly beautiful example of the remnant of one of these massive explosions is the Crab Nebula (see Figure 3.1).

DOI: 10.1201/9781003152965-3

FIGURE 3.1
Image of the Crab Nebula from NASA's Hubble Space Telescope. Pub. Domain

The supernova creating the Crab occurred in our galaxy in the year 1054 AD, when it was recorded as a "guest star" by Chinese and Japanese astronomers. Although the supernova was visible during daytime for 23 days there were no credible European records of the event because written records about anything during this time were very sparse there.

The Threat from Exploding Stars

To get an idea of the violence of supernova explosions, let us imagine what would happen if our sun were to undergo one. Such an event would cause the temperature of the Earth's surface to rise to about 100,000 °C (180,000 °F), which is 15 times greater than the present surface temperature of the sun! In fact, the explosion would, in a short time, vaporize the entire Earth. A solar supernova might first drive any unvaporized astrophysicists insane, since their theories say such an event is not possible. Our sun is just not massive enough for its self-gravity to cause the collapse of its core that triggers the supernova. Still, even if a solar supernova is impossible, a massive star exploding sufficiently nearby could still do us in by destroying Earth's protective ozone layer.

Stars massive enough to become supernovae are also extremely luminous because they burn their nuclear fuel at a prodigious rate. As a result, such stars burn out very quickly, and therefore are very rare at any one time. Some particularly massive stars, known as *supergiants*, can have lifetimes as short as a few hundred thousand years, which is roughly 50,000 times less than our sun's estimated lifetime. In human terms, such "fast-living" supergiants would be equivalent to a species of people who lived their entire life in a single day. Fortunately, given their rarity, no massive stars are close enough to threaten all life on Earth if they became a supernova. The two nearest massive stars likely to become supernovae sometime within the next 10,000 years or so are the supergiants Antares and Betelgeuse.

Antares is the brightest star in the constellation Scorpius, and Betelgeuse (pronounced "beetle juice") is one of the two brightest stars in the constellation Orion. In Greek mythology, Orion was a hunter, as well as a giant, and Betelgeuse is located near the giant's left shoulder. In fact, the name "Betelgeuse" derives from the Arabic, and means the hand or the armpit of Orion. To get an idea of the star's massive size, imagine its center were located at the center of the sun. In that case, the orbits of Mercury, Venus, Earth, Mars, and Jupiter would all lie beneath the star's surface. Fortunately, Betelgeuse's distance (700 light-years) places it well beyond the estimated 100 light-year distance, from within which a supernova explosion would be likely to destroy Earth's protective ozone layer.

If Betelgeuse were to become a supernova, it would become the brightest one ever witnessed by humans, reaching a brightness as great as the full moon, and making it easily visible during broad daylight. While such an event would pose no physical danger to Earth, it surely would be interpreted in apocalyptic terms by many people. This possibility is likely to be especially acute during a time of great global anxiety. Sociological consequences aside, however, you can take supernovae off your "things-to-be-scared-of" list.

The Birth, Life, and Death of Stars

Stars are born when the denser regions in huge clouds of interstellar dust and gas collapse under their own gravity, and the resulting compression makes these regions dense and hot enough (about 10 million °C) to cause the hydrogen nuclei they contain to fuse or join. Nuclear fusion differs drastically from the ordinary burning of hydrogen gas, which involves a rearrangement of electrons in hydrogen atoms. In contrast, fusion involves the combination of light atomic nuclei to form heavier ones. An important example is the multistep process in which four hydrogen nuclei combine to form one helium nucleus. Such a hydrogen fusion reaction releases a million times greater energy than the ordinary burning of hydrogen gas and it requires a much higher temperature to initiate.

For most of a star's life, the outward pressure due to the outflow of heat energy from hydrogen fusion in its core just balances the inward pressure due to gravity. During this relatively stable phase of its life, the star gradually sheds mass, and emits energy as it converts more and more of the hydrogen in its core into helium. Our sun, for example, produces the energy it emits by losing around 4 million tons of mass every second due to fusion. According to Einstein's $E = mc^2$, this mass loss is equivalent to the energy emitted by about 100 billion hydrogen bombs detonating in the sun's core every second – and that's during the sun's stable phase!

Hydrogen is the lightest element, but it is by no means the only light element that undergoes fusion in stars. As the core's hydrogen becomes depleted and its helium content increases, our sun and other stars that are sufficiently massive will, at some point, and heat up enough to fuse their core's helium, thereby producing carbon. Helium fusion begins after the depletion of hydrogen in the core causes a reduced rate of hydrogen fusion, and a drop in outward pressure. That drop in outward pressure leads to a contraction of the core, and an associated rise in its temperature to about 100 million °C. Our sun, however, does not have enough mass to take the next step of fusing carbon nuclei, which would require a temperature of 600 million °C. As this example illustrates, each step in this process of "cooking up" progressively heavier atomic nuclei occur at a higher temperature generated by a partial core collapse. The full chain of fusion reactions ends with the fusion of silicon to iron, which requires a temperature of 3 billion ° C. To reach that last step a star would need to have at least eight times the sun's mass. As Figure 3.2 shows, the core of the star when this last step is reached has an onion-like structure, with the innermost rings fusing the heavier elements at higher temperatures and doing so over a shorter time span. If a star reaches this onion-like structure with iron at the center it has reached the "end of the road," and its cataclysmic death is imminent.

The reason the last step in the sequence of fusions has iron nuclei at the center is because fusion of iron (or any heavier nuclei) does not produce energy but rather

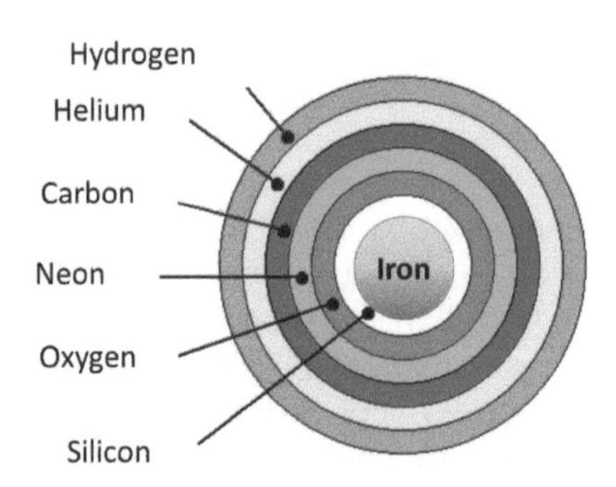

FIGURE 3.2
Core of a massive star showing shells of successively heavier elements toward the center.

consumes it. The resulting loss in energy production when iron nuclei are at the center of the core causes a sudden loss of outward pressure, and hence a rapid collapse of the core. This violent *implosion* then triggers a gigantic thermonuclear explosion of the unconsumed hydrogen in the outer layers of the star. In contrast to this cataclysmic end, lighter stars like our sun, while they will undergo dramatic changes, eventually go out with a whimper rather than a bang. The final state of our sun will be a "black dwarf," (not to be confused with a black hole), which is essentially a burnt-out cinder of a star that no longer shines.

When supernovae occur, they briefly outshine the whole galaxy in which they occur, and they radiate more energy than our sun will in its entire lifetime, which is equivalent to roughly 100 billion H bombs exploding each second for around 10 billion years! Even more remarkably, the unimaginable amount of light and heat that supernovae emit comprise less than 1% of the total energy emitted, *with the remaining 99% in the form of neutrinos and antineutrinos!*

Remarkably, the two words *neutrino* and *SN 1987A* appear in over 19,000 scientific papers since that supernova occurred. This intense interest has arisen because SN 1987A was the first and only supernova observed to emit neutrinos. Until 1987, no neutrinos had ever been observed coming from any astronomical object other than the sun. Even now, although more than ten thousand supernovae have been observed optically in other galaxies, neutrinos have not been detected coming from any of them. We know the neutrinos are present from extragalactic supernovae, but they just have not been observed, owing to the great distance to other galaxies – there are just too few of them reaching us from each supernova to distinguish them from other background sources.

BE THE FIRST TO SPOT A NEW SUPERNOVA

The SuperNova Early Warning System (SNEWS) is a network of neutrino detectors designed to give early warning to astronomers, including amateurs, in the event of a supernova in our own galaxy. Since most neutrinos leave the supernova without interacting, they would reach Earth before the light which has to "struggle" to leave. Thus, the point of the warning system is to allow astronomers to start observing the precursor star before the supernova is detected visually. SNEWS is not capable of detecting most extragalactic supernovae, but many of these have been first spotted optically by amateurs by comparing images of a starfield taken at different times. In 2011, for example, Canadian Kathryn Gray, then 10 years old, became the youngest person to ever discover a supernova, but she shares credit for the discovery with amateur astronomers Dave Lane and her father, Paul Gray. In a case of "extragalactic sibling rivalry," Kathryn's discovery was partially eclipsed by that of her brother Nathan, who two years later discovered his first supernova also at age 10, but he was 33 days younger than Kathryn.

How to Detect Neutrinos

There were four neutrino detectors operating on February 24, 1987, when SN 1987A was observed, and the largest of these was Kamiokande-II (K-II) under Mount Ikeno in Japan. That detector is no longer operating, but its far larger and more sensitive descendant is known as Super-Kamiokande, and its interior is depicted in Figure 3.3. The enormous size of this detector can be judged based on the image of a boat near the center. People in the boat are inspecting some of the 11,000 photo-sensors that line the walls. Those sensors do not detect neutrinos directly. Instead, when the detector is entirely filled with 50,000 tons of very pure water, a charged particle created by a neutrino interaction in the water generates a shock wave in the form of a cone of light – see Figure 3.4. That cone activates some of the photo-sensors forming a circular ring of "hits." Based on the pattern and number of these hits, the detector can determine the energy of the neutrino that caused them. Of course, this cone of light has nothing to do with the light cones in the previous chapter!

An *event* in particle physics often refers to the result of an interaction that takes place between subatomic particles. A major concern of neutrino experimenters is that of background processes which can mimic events induced by neutrinos, such as cosmic rays, and radioactivity inside or near the detector. In designing a neutrino experiment many precautions must be taken to reduce the background, including putting the detector deep underground (to

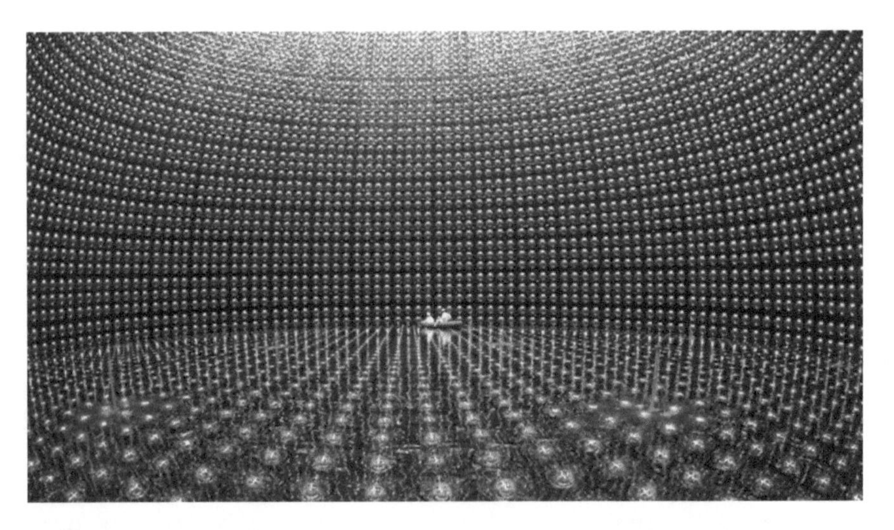

FIGURE 3.3
The Super-Kamiokande detector is located 1,000 meters below ground in a mine near the Japanese city of Kamioka.

Source: Kamioka Observatory, ICRR, The University of Tokyo/ NHK Enterprises Inc.

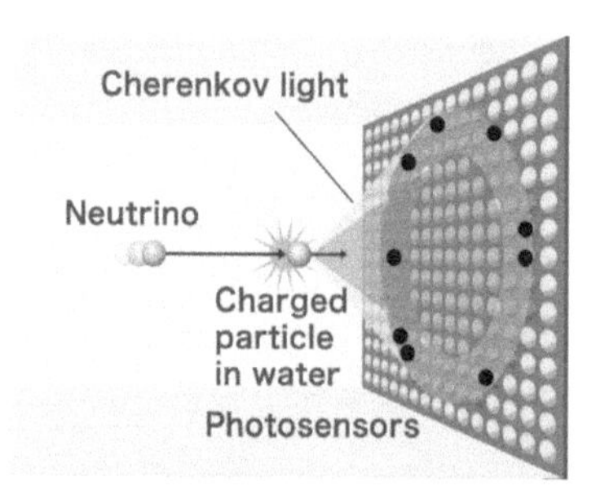

FIGURE 3.4
Conical shock wave of light incident on photo-sensors shown activating nine of them.

filter out cosmic rays), using very pure water, and only counting events which are not too close to the detector walls, where outside radioactivity becomes more of a concern. However, even with such precautions, most of the time that the detector is activated it will be by a background event. Only when a significant number of photo-sensors are activated (many "hits"), all occurring during a very short time interval, is it more likely that the cause is an incident neutrino. Even then, the neutrino may be from a source other than the one they wish to observe. The observed quantities that define an event are (a) the time it occurred, (b) the number of hits or photo-sensors activated, and (c) the very rough direction from which the neutrino came. On February 23, 1987, the neutrino detectors then operating registered something never observed before or since, namely the near simultaneous arrivals of multiple neutrinos, that is neutrino *bursts*. These bursts were recorded within hours of an observed supernova seen in the Dorado constellation.

The SN 1987A Neutrino Burst

Figure 3.5 shows the data for the neutrino burst recorded by the Kamiokande-II detector on February 23, 1987. The coordinates of each dot in the figure shows time on the horizontal axis, and how many sensors were hit at that time on the vertical. The time labeled t = 0 corresponds to 7:35:35 World Standard Time, which is indicated by the faint vertical line. Within seconds of t = 0, the figure shows there were a dozen events, most having more than 20 hits. As can be seen, events having at least 20 hits are almost completely absent at earlier and later times. The number of hits is related to the neutrino energy, and 20 hits correspond to an energy of about 8 MeV. Thus, as can be seen from Figure 3.5, to be easily distinguished from random background events, the neutrinos in this burst needed at least 8 MeV of energy. The association of the neutrino burst with the star Sanduleak-69 202 was confirmed when

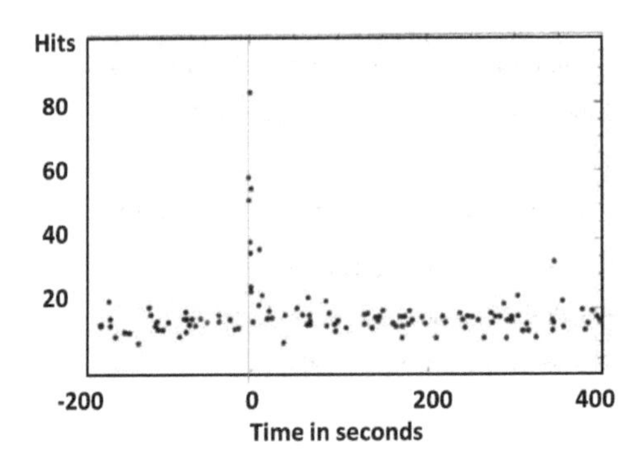

FIGURE 3.5
Plot showing about 100 events observed during a 600 second interval on February 23, 1987 in the K-II detector. The time t = 0 is 7:35:35 World Standard Time. Data extracted from K. S. Hirata et al., Physical Review D, Volume 38, no. 2, 1988.

a sudden abrupt growth in that star's brightness was observed a few hours after t = 0, making it newly visible to the naked eye. The arrival of the neutrino burst *before* the star increased its brightness seemingly suggests that the neutrinos traveled faster than light (FTL), but such a conclusion is unwarranted. The neutrino burst reached Earth before the light from the supernova because unlike neutrinos, photons of light are delayed by many interactions with matter before they can escape the stellar core.

The number of neutrinos emitted from a supernova is about 10^{58}. One indication of the incredible size of this number is that it is about 10 trillion trillion trillion times the number of stars in the whole universe, or about 100 million times greater than the number of atoms making up the entire Earth. Yet of those 10^{58} emitted neutrinos a mere 12 were detected by the Kamiokande-II detector. Even fewer neutrinos were recorded in the other three detectors operating in 1987 due to their smaller size. Those other detectors with their locations and the numbers of neutrinos they recorded were: IMB in the U.S. (8), Baksan in Russia (5), and LSD in Italy (5). LSD is also known as the Mont Blanc detector, named for its location under the highest mountain in the Alps. The neutrino bursts seen in Baksan and IMB occurred within seconds of that in Kamiokande-II, as would be expected. Crazily, the burst seen in the Mont Blanc detector was observed nearly five hours early compared to those in the other three detectors.

If we assume the neutrinos seen in the other three detectors traveled at essentially light-speed, the 5-hour early arrival of the Mont Blanc neutrinos seemingly suggests they had a speed slightly greater than light. The actual excess would be equivalent to beating the nearest competitor in a 26.2-mile marathon by about the width of a human hair. However, any excess above light speed, no matter how tiny, is anathema to most physicists. Therefore, the five-hour early arrival of the Mont Blanc neutrinos has led most physicists to conclude that this burst was just a weird unexplained occurrence having no relation to SN 1987A, an opinion I shared until recent years. As we shall see later, however, the strange Mont Blanc neutrino burst plays an essential role in uncovering the evidence for our three unicorns.

SN 1987A and the 3 + 3 Model of Neutrino Masses

This section describes the origin of my 3+3 model of the neutrino masses. In contrast to the initial sections of this chapter, which have dealt with commonly accepted facts and theories about supernovae, all the rest of it deals with this controversial model. As the model's originator, I believe the model is correct, of course, even though most neutrino physicists think it cannot be right, and that I am hopelessly deluded. The model (see Figure 3.6) differs significantly from the conventional picture of the neutrino masses depicted in Figure 1.11 of Chapter 1. It is called 3+3 because it postulates three sterile neutrinos each paired with an active one having nearly the same mass. The separations between the mass square (m^2) of the members of each active-sterile pair, that is the dm^2 values, are also shown in the figure. These dm^2 values are based on the wavelengths observed for neutrino oscillations, which are here assumed to be between active and sterile neutrinos. The dm^2 values for the two positive m^2 active-sterile pairs are well established but that for the third pair, $dm^2 \approx 1\ eV^2$, is only approximate, being based only on a few experiments. Furthermore, those experiments are controversial, because a value as large as $dm^2 \approx 1\ eV^2$ would imply the conventional neutrino mass picture in Figure 1.11 cannot be correct, given the small separations shown between the three masses and the lack of sterile neutrinos. In fact, it was the ***non***existence of any oscillations having a third dm^2 besides the two depicted in Figure 1.11 which gave rise to that conventional picture in the first place. The 3 + 3 model is not the only one postulating three sterile neutrinos, but it

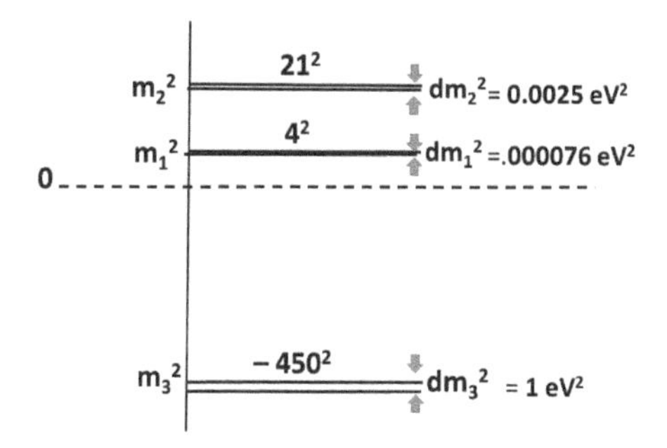

FIGURE 3.6
The 3 + 3 model of the squares of the neutrino masses. This model includes 3 active-sterile pairs, one of which is a tachyon with negative m^2. Not drawn to scale.

is unique in specifying actual values for the three masses, and most controversially, the inclusion of one active-sterile neutrino pair having a negative m^2, that is, a tachyon.

Let us see how the 3 + 3 model came about, based on an analysis of the few dozen neutrinos observed from supernova SN 1987A. To begin, we note that the arrival times of those neutrinos in three of the detectors (all except Mont Blanc) were spread over about 15 seconds. In the usual analyses of these data, it is assumed that spread in arrival times must reflect the spread in the time of neutrino *emission* from SN 1987A. The basis of that assumption is that if the neutrino masses are all nearly equal, as the conventional model assumes, the spread in their travel times from the star would be completely negligible. Therefore, the spread in arrival times must be the result of varying emission times. In contrast, the 3+3 model was obtained by making no assumption about the near equality of the neutrino masses. Instead, the model assumes that the neutrinos are mostly emitted from the star during a short time interval lasting perhaps 2 seconds, and the spread in the neutrino arrival times mainly reflects their varying *travel* times. Given this assumption, we can calculate the mass of *individual* neutrinos making up the SN 1987A burst based on their observed energy and arrival time. When the analysis is done, we come to the striking conclusion that the observed few dozen neutrinos all had masses consistent with being either: $m_1 = 4$ eV or $m_2 = 21$ eV within uncertainties. By the way, an eV, short for an *electronvolt*, is a unit of energy not mass. Thus, by $E = mc^2$, the two preceding masses really should have been written as 4 eV/c^2 and 21 eV/c^2 but we will not worry about the omission of the c^2.

Figure 3.7 shows the computed mass for each neutrino based on its measured energy and arrival time, with an uncertainty depicted as a horizontal error bar. As seen from the figure, the neutrinos are not found to have masses of exactly 4 eV and 21 eV, of course, but they merely cluster around those two values. Such departures from exact values would be expected, given measurement uncertainties and unsynchronized detectors. Neutrino masses of 4 eV and 21 eV are tiny when compared to those of any other particle, but they are impossibly far apart based on the conventional neutrino mass model. In that conventional model, the three neutrinos have m^2 values separated by a mere 0.0025 eV^2 as shown in Figure 1.11 of Chapter 1.

After I found the puzzling "two-mass clustering" result, I wrote a paper on it in 2012. My discovery was in fact only a ***re***discovery of something pointed out 24 years earlier by Ramanath Cowsik, and a year before him by Humiaki Huzita! Before submitting my paper for publication, I contacted Cowsik for feedback. Cowsik noted that I had added some important ideas to the strange clustering about two mass values for the SN 1987A neutrinos that he (and Huzita before him) had earlier observed. However, he expressed the view that the clustering about two masses as widely separated as 4 eV and 21 eV had to be just a chance occurrence, because it was inconsistent with the widely accepted neutrino model. My reaction to the strange result, however, was very different than Cowsik's. I began to doubt the correctness of the

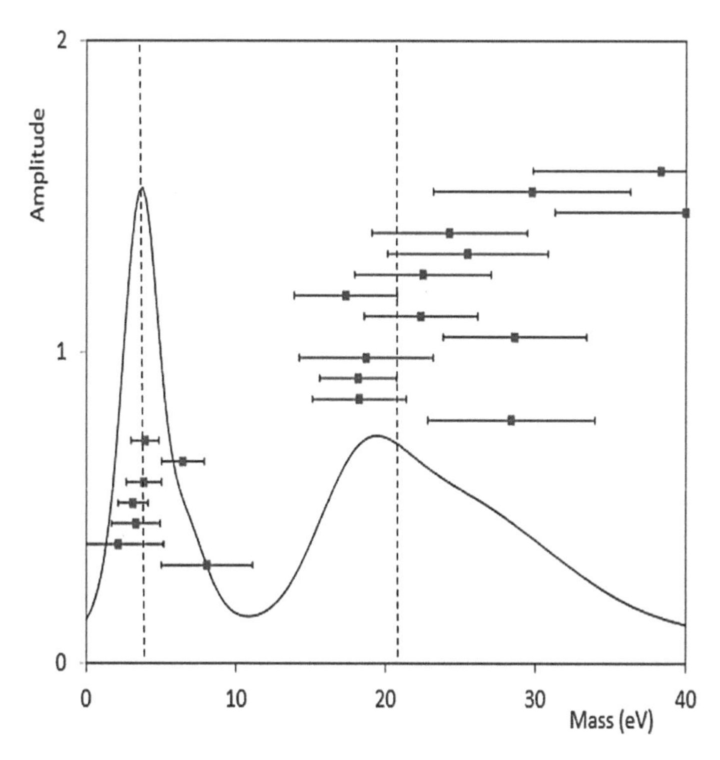

FIGURE 3.7
Clustering of the neutrino masses about 4 eV and 21 eV (vertical lines). The curve is found by adding 20 Gaussian distributions – one for each neutrino.

accepted model, and I proposed the 3+3 model the following year. Even though only two masses were observed, this model needed to include three active-sterile neutrino pairs, because there is very good evidence that there are three active neutrinos.

The Tachyonic Third Mass in the Model

The most controversial element of the 3+3 model is the tachyonic mass of the third neutrino pair, namely $m_3^2 = -450^2$ eV^2, which can also be written as $m_3^2 \approx -0.2$ keV^2. The basis of this choice of m_3^2 for the third pair was the striking coincidence that the 4 eV and 21 eV pairs were found to have the same *fractional* mass separation, that is, the same value of the quantity dm^2/m^2 of five parts per million. If this equality was more than a weird coincidence, it was therefore natural to assume the same value held for the third neutrino pair. Thus, if we use the approximate value $dm_3^2 \approx 1$ eV^2 from several oscillation experiments, and assume the same dm^2/m^2, we obtain $m_3^2 = -450^2\ eV^2$. Clearly, the 3+3 model was not the result of any elegant theory. Instead, it was based on (a) the observed clustering of the SN 1987A neutrino masses about two specific values, 4 eV and 21 eV, and (b) an observed odd numerical coincidence, namely the observed equality of the two values of dm^2/m^2.

A careful reader may wonder why the third mass in the 3+3 model was assumed to have a *negative* m^2, since we could find the same magnitude fractional mass separation (the same dm^2/m^2) had we used a positive m^2. To understand the basis of the choice of a negative m^2 for the third neutrino mass, we first need to define the *effective mass*. The value of an effective mass is defined by having its *square* to be a weighted average of the squares of the three individual masses m_1, m_2, and m_3 with specific weighting factors. By this definition, given the positive values of m_1^2 and m_2^2, if we want the effective mass squared to be negative, this can only happen if only if m_3^2 is negative. Finally, you might ask why does the effective mass squared need to be a small negative value? Recall that in Chapter 2, it was found that a small negative m^2 for the electron flavor neutrino could explain the *knee* in the cosmic ray spectrum. This feature of the spectrum was assumed to be the result of proton beta decay being allowed above some threshold energy – something only possible for an $m^2 < 0$ (tachyonic) effective mass for the electron flavor neutrino.

Balancing a Seesaw with Nothing on One Side

The seesaw analogy depicted in Figure 3.8 may clarify why the choice of a negative m_3^2 allows the effective mass squared to be both negative and very small, when the other two masses are neither. The figure shows three masses on one side of a seesaw, one of which has a negative weight – say in the form of a large helium balloon for m_3. If the three distances from the pivot

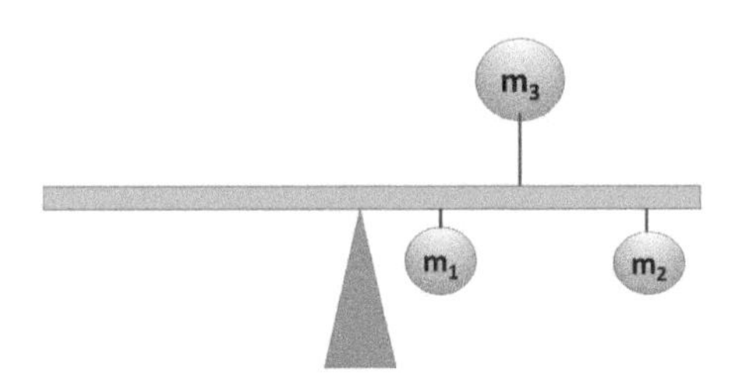

FIGURE 3.8
A seesaw analogy to explain why the third mass pair is chosen to have a negative m^2 in the 3 + 3 model. The mass m_3 is a helium balloon.

are chosen properly, the effective weight of all three masses could be zero, meaning that the seesaw could balance with nothing on the other side. With a slightly larger helium balloon the effective weight on the right side of the seesaw would be negative and very small. This analogy, of course, is faulty because for neutrinos it is the *square* of m_3 that is assumed to be negative, not m_3 itself. For readers who are neutrino experts, I also note that this seesaw analogy has nothing to do with the so-called seesaw mechanism, which is invoked to explain the extreme lightness of the neutrino masses compared to all other particles.

When the 3+3 model was first proposed in 2013, I considered it extremely speculative, so much so that I probably was lucky to have it published in a reputable journal. Not only did I claim some neutrinos were tachyons, that is, had a negative m^2, but I based the model largely on an observed numerical coincidence, rather than some elegant new theory. In addition, the model rested on a shaky foundation because the statistical evidence for the SN 1987A neutrinos clustering about two mass values (4 and 21 eV) was less than rock solid, resting as it did on just a few dozen neutrinos. With all the reservations I had about my "baby," I knew that it would not long survive unless I could produce some corroborating evidence. On the other hand, I was also aware that having a highly speculative model that few other physicists took seriously was a distinct advantage. It was much less likely that other researchers would be on the lookout for confirming evidence or even be aware of my model's existence.

Seeking Validation for the 3 + 3 Model

If another galactic supernova were to occur, it would surely reveal if the three neutrino masses had the values stipulated in the 3+3 model, given the far greater sensitivity and size of today's neutrino detectors compared to those in 1987. A galactic supernova today might result in thousands of observed neutrinos rather than the few dozen seen in 1987. I initially foolishly believed that since the 3+3 model was published in 2013, and the last supernova occurred 26 years earlier, we were due for another soon. Regrettably, I had fallen victim to the "gambler's fallacy" – the mistaken belief that the occurrence of past, randomly distributed events affects the likelihood of future events. In fact, a long string of years with no supernova (or a coin landing heads) does *not* make it more likely that a supernova will soon occur (or that a coin will land tails). Instead of waiting for the next galactic supernova, a prudent 83-year-old physicist should seek validation of his model elsewhere, since his chance of seeing the next one is miniscule. Much to my delight this corroborating evidence has steadily accumulated, beginning with some results involving dark matter.

Dark Matter Holds Galaxies Together

The Milky Way Galaxy, which contains our solar system, is in the form of a disk about 100,000 light-years in diameter and 10,000 light-years thick. The galaxy does not, of course, rotate as a solid disk, but rather each star is in its own orbit about the galactic center. Since our solar system is in the plane of the galaxy, we see part of the galaxy as a beautiful band across the sky that looks like dust but is really many individual stars (about 250 billion of them) that cannot be distinguished as individual stars by the naked eye – see Figure 3.9. Nowadays, it is only in remote areas without light (or other) pollution that we can see this wondrous sight. Until the early 1920s, most astronomers thought that the Milky Way contained all the stars in the Universe, but we now know there are roughly as many galaxies in the universe as there are stars in our own.

FIGURE 3.9
The Milky Way. Photo: Milky Way as seen from Atacama Desert of Chile.

Source: ESO. P. Horálek

The hypothesis of dark matter has a long history going back to 1884 when Lord Kelvin (aka William Thomson) suggested that "many of our stars, perhaps a great majority of them, may be dark bodies."[2] Another key figure was Swiss astrophysicist Fritz Zwicky who suggested in the 1930s that dark matter needed to be present in galaxy clusters to hold them together. However, it was not until the 1970s that observations by Vera Rubin (see Figure 3.10) and her colleague Ken Ford led to their discovering the solution of the "galaxy rotation problem." According to Rubin and Ford, the Milky Way and virtually all spiral galaxies are spinning way too fast to be held together by the gravitational pull of the visible matter observed in telescopes. Rubin and Ford reasoned there had to be a giant sphere of invisible (dark) matter in galaxies that allows them to rotate as rapidly they are without flying apart. In fact, their observations showed that dark matter forms a spherical halo much larger than the diameter of the disk of visible matter (stars, dust, and gas) in the galaxy.

FIGURE 3.10
Vera Rubin when she was an astronomy major at Vassar College.

Source: Archives and special collections, Vassar College Library. Ph.F 6.93 Cooper Rubin

First Confirmation of the 3 + 3 Model Masses

The nature of dark matter remains a mystery, but sterile neutrinos have often been considered as a promising candidate. In 2014, I serendipitously learned of some dark matter simulations done by Man Ho Chan, a young Chinese astrophysicist residing in Hong Kong. After some correspondence between us, Man Ho and I began collaborating on a new simulation, assuming the dark matter in the galaxy consists of sterile neutrinos. We found, much to our delight, that the orbital velocities of stars in the Milky Way Galaxy versus their distance to the center could be nicely explained by assuming a spherical halo of sterile neutrinos, but *only* if the neutrino mass was very close to 21 eV, which is one of the two positive m^2 masses in the $3+3$ model.

But there was more. Galaxies typically hang out together, and often belong to large clusters consisting of many thousands of individuals. Man Ho and I concluded that sterile neutrinos could keep galaxy clusters from flying apart, but only if their mass was close to 4 eV, which is the other positive m^2 mass in the $3+3$ model. Of course, it is likely both 4 eV and 21 eV neutrinos would be present to some degree in both individual galaxies and their clusters. Nevertheless, it seemed reasonable that the smaller of the two masses would be the main one found in the larger of the two structures, that is, clusters versus individual galaxies. The reason smaller masses favor larger structures can be understood if a mixture of the two neutrino types was initially present at a common temperature. In that case, the lighter faster moving neutrinos would then have higher speeds and they would more easily escape the individual galaxies into their surrounding clusters. Having found evidence from our dark matter simulations supporting the values of the $m^2 > 0$ masses in the 3 + 3 model, it was natural to wonder where one might look to find evidence for the third $m^2 < 0$ (tachyonic) mass?

Revisiting the Mont Blanc Neutrino Burst

As noted earlier, nearly all physicists have dismissed the nearly 5-hour early neutrino burst in the Mont Blanc detector, as being a random occurrence unrelated to SN 1987A, since it implied that the neutrinos in that burst were slightly superluminal or FTL. It may seem strange, but I initially shared this mainstream view that the Mont Blanc burst was unrelated to SN 1987A, even though I had been looking for evidence that some neutrinos were tachyons! Recall that this burst arrived nearly 5 hours (or more precisely 16,900 seconds) early during a 7-second time interval. If the burst were genuine, the individual neutrinos in the brief burst would have a spread in their arrival times relative to that of light of 7 parts in 16,900 or a mere 0.05%. Thus, if these five neutrinos had a common imaginary mass, they would necessarily have a miniscule spread in their energy. In effect, the Mont Blanc neutrinos would constitute a *line* in the spectrum, or equivalently, they had to be *monochromatic*, which literally means "one color," or one energy.

Could the 5 neutrinos comprising the Mont Blanc burst, whose average energy was 8 MeV, truly be monochromatic? It would be nice if one could measure their energies very precisely to see if this were so, but the large measurement uncertainty makes that impossible with any precision. Still, it is very striking that the energies of the five Mont Blanc neutrinos are all equal within their ±25% measurement uncertainty, which is quite unlike the neutrino bursts seen in the other three detectors. Thus, the neutrinos in this burst are at least *consistent* with being monochromatic. These facts make the Mont Blanc neutrinos very different from those observed in the other three detectors. In particular, the SN 1987A neutrinos observed in the other detectors had energies that were (a) much higher than 8 MeV, and (b) not monochromatic, given the large spread in their energies. Finally, there is a third coincidence with respect to this neutrino burst. Based on their five-hour early arrival, if the five Mont Blanc neutrinos were assumed to have a common mass, its value was, within a factor of two, equal to that which I had previously stipulated for the tachyons in the 3 + 3 model.

Despite these *three* odd coincidences, I still initially rejected the possibility that the Mont Blanc neutrinos were the tachyons in the model. Clearly, I had been too influenced by supernova modelers who assured me that an 8 MeV neutrino line simply could not be part of the supernova spectrum. In fact, I even foolishly stated in my 2013 paper introducing the 3 + 3 model that having an 8 MeV line in the SN 1987A spectrum was *inconceivable for neutrinos from a supernova.*[3] I have now learned the hard way never to use the word "inconceivable" in a scientific paper. Five years after proposing the 3 + 3 model, I *conceived* of a way to obtain 8 MeV monochromatic neutrinos from a supernova, which I had formerly considered to be inconceivable!

The Neutrino Detector That Failed to "Bark"

Before considering how to obtain monochromatic neutrinos, let us deal with one troubling aspect of the Mont Blanc burst that has made many physicists reluctant to assume it was genuine, aside from its 5-hour early arrival. Why was no such 5-hour early burst seen in the other three detectors – especially the Kamiokande-II (K-II) detector, the largest of them all? This absence of a 5-hour early burst in K-II has convinced most physicists that the Mont Blanc burst could not be real. This "missing burst" in the K-II detector recalls a famous Sherlock Holmes murder mystery in which the guard dog at an estate did not bark when the killer entered the grounds. That absence of a bark suggested to the great detective that the dog must have known the killer.

As in the Sherlock Holmes story, there was something special about both the Mont Blanc detector and the nature of the neutrino burst it recorded, which explains why a 5-hour early neutrino burst would not cause the K-II detector to "bark" when it arrived. As already noted, the neutrinos in the Mont Blanc burst all had energies close to 8 MeV – a much lower energy than the neutrinos seen in the main burst at $t = 0$ in the other three detectors. Since the Mont Blanc detector had an energy threshold much lower than the other three, it could observe 8 MeV neutrinos much more easily than the others, in which they were hidden by the background events. In fact, the K-II detector had only a 20% detection efficiency for 8 MeV neutrinos compared to those of higher energy. In addition, the four detectors were not well-synchronized to universal time, being off by perhaps plus or minus one minute. This means that we are not looking for signs of a burst in K-II at a very precise time, but rather somewhere in a two-minute time interval, making it that much harder to spot, given a sizable background. These two facts make it easy to understand how a small 5-hour early burst of 8 MeV neutrinos might have been hidden by the background events in the K-II detector.

Dark Matter in the Stellar Core and 8 MeV Neutrinos

Assuming the Mont Blanc neutrino burst really did consist of 8 MeV monochromatic neutrinos, let us now see what sort of process in the supernova might have created them. Just as dark matter is present in galaxies and their clusters, it might also be present in the cores of massive stars. The dark matter particles in a stellar core might well have been part of the huge cloud of dust and gas which collapsed to form the star, with the dark matter helping to facilitate the collapse. In 2018, I proposed a dark matter model explaining how to create 8 MeV monochromatic neutrinos from a supernova, based on dark matter X-particles of mass 8 MeV in the core of massive stars. This value for the X mass was not arbitrary rather it was based on a new particle (Z′) that Attilla Krasznahorkay and his Hungarian colleagues at the Atomki Lab in Hungary had then recently discovered in a nuclear physics experiment. This Z′ particle has a mass of 16.7 MeV, and apparently rapidly decayed into electron-positron pairs. After this discovery Jonathan Feng and his collaborators identified the Z′ to be a *mediator* between dark matter and normal matter. Feng had essentially proposed that the Z′ particle was a *portal* linking the world of dark matter X-particles to ordinary particles (like electrons and neutrinos). There was nothing mystical about Feng's portal, which could also be called a "Z′-mediated reaction" (see Figure 3.11).

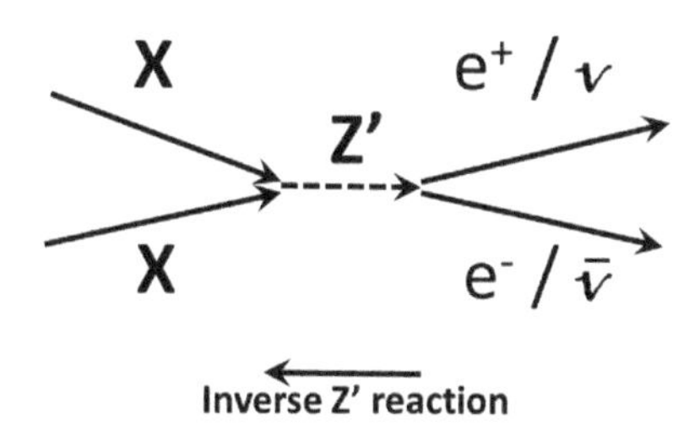

FIGURE 3.11
Z′-mediated reaction and its inverse connecting dark matter X-particles and either neutrinos or e^+ + e pairs.

My model for creating an 8 MeV neutrino line from SN 1987A built on Feng's idea by assuming that during a supernova, pairs of X-particles meet up and annihilate. The XX annihilation then briefly forms a Z′, which rapidly decays into a pair of leptons, that is, either an electron and positron or a neutrino and antineutrino. This Z′-mediated reaction releases about 16 MeV of energy, based $E = mc^2$ and on the mass difference between the initial and final particles. However, the reaction occurs only when some X-particle ignition temperature is reached in the stellar core. In Feng's terminology, the portal between the worlds of dark matter and ordinary matter in a stellar core is normally closed, but it opens if the temperature on one side of it is high enough. You can also think of the situation as if the X-particles in the stellar core were an inert pile of kindling (see Figure 3.12), which when ignited, results in an enormous energy release in the form of much lighter $e^+ e^-$ pairs or neutrino-antineutrino pairs.

The dark matter X-particles are assumed to be initially "cold," meaning they are traveling with velocities much smaller than light. They are cold because dark matter feels only the force of gravity, and hence it does not take

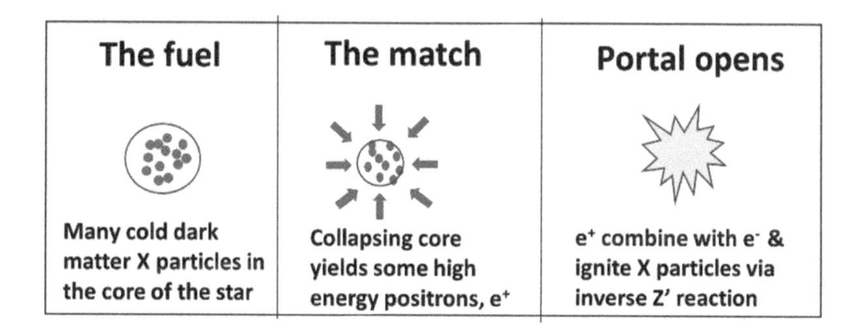

FIGURE 3.12
The process by which 8 MeV dark matter X-particles in the star's core combine to release energy in the form of 8 MeV $e^+ e^-$ and neutrino- antineutrino pairs by the Z′-mediated reaction.

part in nuclear fusion reactions in the star's core prior to it becoming a supernova. As a result, when two slowly moving X-particles in the core combine to briefly form a Z′ (see Figure 3.11) it would therefore be created nearly at rest. In that case, energy conservation requires that the emitted neutrinos and antineutrinos from Z′ decay would each have an energy equal to half the Z′ particle's rest energy, or $E = \frac{1}{2}(16.7) \approx 8$ MeV, and they, therefore, would be very nearly monochromatic.

Support for the Z′-Mediated Reaction

Having proposed the Z′-mediated reaction model in 2018 for creating monochromatic 8 MeV neutrinos from SN 1987A, I considered it urgent to find empirical support for it. Besides having dark matter X-particles in stellar cores, it is logical to assume that some X-particles might also cluster near the center of our galaxy, where a supermassive black hole is known to reside. Such objects can also be found in the cores of most other spiral galaxies. Our galaxy's supermassive black hole corresponds to the location of Sagittarius A*, and it has a mass over four million times that of the sun. This object first revealed its presence when UCLA astrophysicist Andrea Mia Ghez (Figure 3.13) and her group observed that stars near the center of the galaxy had very rapid orbits around some unseen object – a discovery for which she received a share of the 2020 Nobel Prize in physics. Ghez, incidentally, was only the fourth woman to receive the Nobel prize in physics out of a total of 218 recipients as of 2021.

Even before the discovery of that supermassive black hole, astronomers have studied the gamma rays emanating from a small spherical region of gas and dust near the galactic center. This is a violent region where gas and dust are heated to millions of degrees when matter falls into the supermassive black hole.

If we seek evidence for the existence of dark matter X-particles of mass 8 MeV near the supermassive black hole, we need to know how such particles would affect the gamma ray spectrum from the hot gas and dust. Suppose we assume that those gamma rays come from the positrons produced in the Z′-mediated reaction that starts with XX annihilation. The positrons so created would then create gamma rays when they encounter other electrons in the gas and dust near the galactic center.

By conservation of energy, if the cold X-particles have a mass of 8 MeV, the gamma rays they create during XX annihilation will have energies only up to 8 MeV. These gamma rays from the galactic center would be in addition to the background from other sources, and they would therefore show up as a broad bump above background in the gamma ray spectrum. In a 2018 article I showed that exactly such a broad bump was present in the published

FIGURE 3.13
Andrea Ghez in 2021, photo by Annette Buhl.

data, and that its shape was very consistent with theory. Moreover, two other predictions of the Z′-mediated reaction model were found to be consistent with the gamma ray data, namely the observed temperature and size of the spherical region near the galactic center from which the gamma rays emanated. These *three* points of agreement gave significant support for the correctness of the Z′-mediated reaction as the source of gamma rays from the galactic center. They also, therefore, supported the existence of dark matter X particles of mass 8 MeV, which if present in stellar cores, would give rise to 8 MeV monochromatic neutrinos (an 8 MeV neutrino line) from a supernova.

Challenges to the Z′-Mediated Reaction Model

There are some loose ends, however, that raise questions about the validity of the Z′-mediated reaction model. In English, the idiom "what about…?" is used to remind someone or perhaps oneself about something that appears to contradict what was just said. The first example of a "what about" is that if the gamma rays from the galactic center are really the result of the Z′-mediated reaction then in addition to seeing gammas rays we should also see monochromatic 8 MeV neutrino-antineutrino pairs – so why have none been found? The failure to observe any neutrinos from the galactic center is easily explained by the fact that their detection is far more difficult than gamma rays, especially given the large neutrino background. In addition, the ability to identify neutrinos coming from a specific direction in the sky is not very good, since the neutrino direction is typically not measured very well unless the neutrinos have extremely high energy. Thus, the non-observation of an 8 MeV neutrino line from the galactic center does not negate the Z′-mediated reaction model.

A second example of a "what about?" is that if the Z′-mediated reaction explains the presence of monochromatic 8 MeV neutrino-antineutrino pairs from a supernova, that same reaction should also create a burst of electron-positron pairs, so where are the positrons? As in the case of the positrons produced near the galactic center, those created in a supernova will shortly annihilate when they encounter nearby electrons and produce pairs of gamma rays. *Gamma-ray bursts* (GRBs) are a well-known phenomenon in astrophysics that have been observed in distant galaxies – see Figure 3.14. They are the brightest events known to occur in the whole universe, and they can last from ten milliseconds to several hours, with an average time measured in seconds. It seems reasonable that GRBs (known to be associated with extragalactic supernovae) are the result of the Z′-mediated reaction leading to a burst of electron-positron pairs.

A third example of a "what about?" is that if there really was an 8 MeV line in the spectrum of SN 1987A neutrinos, we would expect it to be present in the data for all three neutrino masses, not just those neutrinos having an

FIGURE 3.14
Artist's illustration of a bright gamma-ray burst.

Source: NASA/Swift/Mary Pat Hrybyk-Keith and John Jones. Public Domain

imaginary mass (i.e., the 5-hour early Mont Blanc burst). Furthermore, we would expect that any monochromatic component of the spectrum that is due to a dark matter process would have a long-lasting emission as well as a brief burst. The reason for the monochromatic component being long-lasting is that dark matter can gain or shed heat much more slowly than ordinary matter. However, this third "what about?" takes quite a bit of explanation and is discussed in the next section.

We conclude the present section by summarizing the three main points so far. First, the Mont Blanc neutrinos had to be monochromatic with energy 8 MeV for them to be the tachyons having the mass postulated in the 3 + 3 model. Second, a plausible dark matter model exists for creating 8 MeV monochromatic neutrinos from supernova based on a so-called Z′-mediated reaction. Third, the spectrum of gamma rays from the galactic center has provided empirical support for that reaction. According to an important principle known as Bayes Rule, we need to update our degree of belief in an unlikely hypothesis when considering new evidence. The existence of the dark matter Z′-mediated reaction model and the empirical support for it, therefore, transforms the hypothesis of the Mont Blanc burst being the tachyons in the 3 + 3 model from a "far-out crazy idea" to a reasonably plausible one. We now shall see how with further evidence it becomes even more likely.

Finding the Unicorn Hidden in the Background

As was noted in the preceding discussion, if 8 MeV monochromatic neutrinos, that is a neutrino line, were present in the spectrum of SN 1987A neutrinos, we would expect to find it for all four neutrino detectors not just the one at Mont Blanc. Furthermore, since dark matter cools much more slowly than ordinary matter, we should find evidence for the 8 MeV neutrino line over time intervals much longer than the few seconds of the 12-event burst seen in the K-II detector. We should observe it as part of either an afterglow or perhaps a precursor emission lasting many hours.

Fortunately, the K-II collaboration decided to publish data not just for the time interval shown in Figure 3.5, which contains the brief 12-event burst, but also for seven other 17-minute long randomly chosen time intervals in the hours before and after the main burst (at t = 0). Since the events in those 119 minutes showed no obvious indication of any other bursts, Keiko Hirata, Masayuki Nakahata and their K-II colleagues might have simply chosen to omit those seven plots from their publication. Luckily for us, they chose otherwise, so we can use those data to count how often events having various numbers of hits occurred out of the 997 events recorded during those two hours before, after, and during the main burst at t = 0. Since the number of

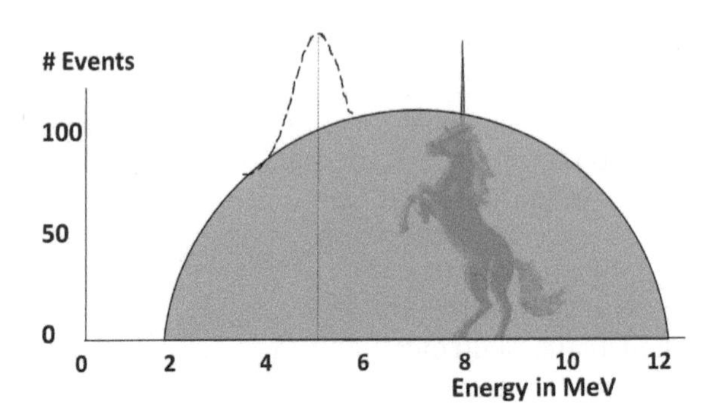

FIGURE 3.15
A fictitious neutrino spectrum showing an 8 MeV line (the unicorn Horn), and a much-broadened line at 5 MeV, both atop a semicircular background.

hits is proportional to the neutrino energy, these events can yield a neutrino spectrum for the hours before and after the main burst, and they allow us to see if an 8 MeV neutrino line has been lurking in these data all these years.

Before looking at what the data show, it is useful to consider a *fictitious* example of what we might expect if there were an 8 MeV neutrino line present in the data. One might imagine a result something like that in Figure 3.15, which shows a background spectrum extending over some range of energies, with a sharp line (the horn of the hidden unicorn) in the number of events at an energy of 8 MeV. Had we been looking at an *optical* spectrum such a result with one or more very sharp lines might easily be observed. Optical spectra having sharp lines are routinely found in the spectrum of light from the sun and other stars, and the wavelengths of the lines tell us what elements are present. Interestingly, the element helium was first discovered in the sun because the position of some observed spectral lines matched no known element. Finding a line in the neutrino spectrum is much more difficult than finding them in optical spectra, however, because of our inability to measure the neutrino energy accurately. In fact, at an energy of 8 MeV the energy of the K-II neutrinos can only be measured to about ±24% or ±2 MeV.

A measurement uncertainty in energy broadens any initially sharp line, and reduces its height by a corresponding amount, while keeping their product constant. The net result is much as if the initially sharp line melted like an icicle standing on end into a puddle. Given the initial height of the sharp 8 MeV line depicted in Figure 3.15, the broadened line would therefore shrink so much in height as to essentially "melt into the background," making its presence virtually undetectable. Clearly, finding evidence for a neutrino line is much more challenging than looking for the very sharp lines seen in optical spectra. Finding neutrino lines also requires having good information about the shape of the background spectrum atop which the line sits.

It is of course possible to find such neutrino lines provided they are sufficiently prominent or tall. For example, suppose a sharp line existed at

5 MeV in our imagined depicted spectrum in Figure 3.15, and that its original height before broadening were perhaps 50 times taller than the "unicorn horn" shown at 8 MeV. Given those assumptions, then after accounting for the line broadening and corresponding height reduction the detector might show something like the dashed curve in the figure. In this example, the presence of such a 5 MeV line would be much more detectable than was the original wimpy 8 MeV line that would melt into the background like an icicle. The key to being able to identify this dashed curve as a true line in the spectrum is having affirmative answers to six questions:

- Does the line have the right shape, i.e., a bell-shaped Gaussian curve?
- Does the line occur at some previously predicted energy?
- Does it have the right width, given the energy resolution of the detector?
- Do we know the shape of the background curve atop which the line sits?
- Do we know the height of the background curve?
- Do we have good statistical significance?

Let us now consider what the K-II data revealed in their second paper on SN 1987A, which the group published in 1988. As previously noted, this paper contained seven "dot plots" (akin to Figure 3.5) for several hours before and after the main 12-event burst. We can use those seven graphs to count how often events having various numbers of hits occurred out of the events recorded during those two hours, and the resulting distribution is shown in Figure 3.16. Vertical error bars are also shown in the figure indicating the statistical uncertainty for each data point. All data points in the central region of the graph (between 14 and 21 hits) can be seen to lie well above the dark background curve. As noted earlier, the number of hits is proportional to neutrino energy, and so Figure. 3.16 is essentially a spectrum, whose peak energy occurs at 17 hits, which is equivalent to 7.5 ± 0.5 MeV. This means that the spectrum is consistent with an 8 MeV neutrino line sitting atop a background curve that peaks at nearly the same energy.

Reliability of the Background

If the 8 MeV line is real, it is extremely worrisome that the background curve and the neutrino line both peak at nearly the same energy. A skeptic might argue that the background curve in Figure 3.16 is spurious, and that the real background might look very much like the data itself – meaning no evidence for any line! The validity of the evidence for a line, therefore, depends entirely on the reliability of the depicted dark background curve. The best way to be sure we are using a correct background is to find both its shape and height independent of the supernova data itself.

A key observation made by the K-II collaboration in their paper was the *stability* of the observed background over time. Given a constant background in time, we could therefore use data taken by the detector at another time to infer the background on the day of SN 1987A. Fortunately, in 1989 the K-II Collaboration published the results of an unrelated search for solar neutrinos. The 450 days of data-acquisition for this search was mostly many months after SN 1987A, and these data were what I used to define the dark background curve shown in Figure 3.16. It is important to note that the height of the background curve in that figure was determined using that solar data, with no arbitrary adjustment made to give a best fit. There is, however, still a reason to be cautious about using the data taken during the months before and after the date of SN 1987A to find the background on the day of the supernova. To understand this concern, consider a totally unrelated problem of finding the age distribution of a town's residents using a random sample of people interviewed on the street. Obviously, for a college town if you randomly sampled people in the downtown area, you would likely get very different results if the sampling were done on a weekday morning or a Saturday night. By the same token, it would not be surprising if the supernova and

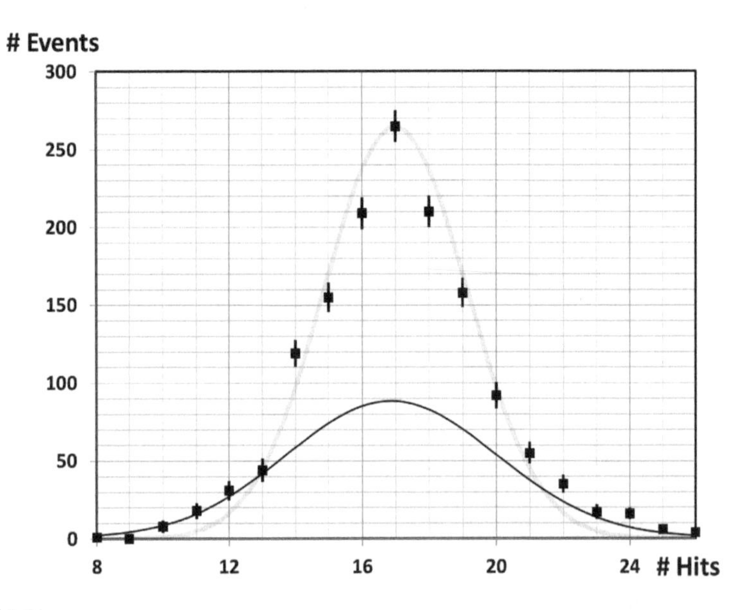

FIGURE 3.16
Distribution of the number of events according to their number of hits for K-II data recorded during 136 minutes on the day of SN 1987A. The dark curve is the background spectrum, found using K-II data taken at a different time.

solar datasets which were taken at different times did have different backgrounds. When those two datasets were recorded there were in fact different "cuts" or selections used.

Thus, for example, the solar data excluded events that were further from the walls of the detector than the supernova data did. Furthermore, the two datasets used slightly different criteria for the minimum number of hits required in a specific short time interval for the detector to be triggered. These differences can be accounted for, however, and they lead only to a small adjustment in the shape and height of the background. In conclusion, differences in the two datasets do *not* prevent us from using the solar data (with small adjustments for the two factors just mentioned) to find the background curve for the supernova data. If, however, we are still concerned that the background height obtained from the solar data is unreliable, there is an alternative. We could simply fit the actual supernova data in Figure 3.16 on the wings of the distribution, that is for $N_{hits} < 14$ and $N_{hits} > 21$ to a simple Gaussian. Such a fit would yield a background curve that is indistinguishable from the one shown.

The 8 MeV Line Revealed

Figure 3.17 shows what the excess above background looks like after subtracting from the height of each data point in Figure 3.16 the height of the background curve in each bin. This figure also has converted the horizontal axis from number of hits to neutrino energy, since these two quantities are directly related. The result in Figure 3.17 can be seen to fit well the familiar bell-shaped Gaussian curve having a central value consistent with an 8 MeV neutrino line. The width of the curve also agrees with the known energy resolution in the detector, which would cause a sharp line to have a spread of ± 2 MeV (measured at half its height). *In summary, the K-II data over several hours on the day of SN 1987A reveal an 8 MeV neutrino line; we have answered all six questions raised in the previous section in the affirmative.*

Assuming this 8 MeV neutrino line in the SN 1987A data is real, it has been very well hidden because the background curve and the line peak at nearly the same energy. But is it possible that this 8 MeV neutrino line is just a mirage? Perhaps it is just a weird chance fluctuation in the background. If the background shape is correct, the probability of a fluctuation as big as the excess counts shown in Figure 3.17, works out to the equivalent of a "30-sigma effect" (or 30 standard deviations). The probability of such a fluctuation is: $p = 10^{-209}$. That is about the same as your chances of flipping an unbiased coin and having it by chance land on heads 694 times in a row. Unfortunately, this probability estimate cannot not be taken too seriously because it depends on the background shape being correct. On the other hand, it is important to

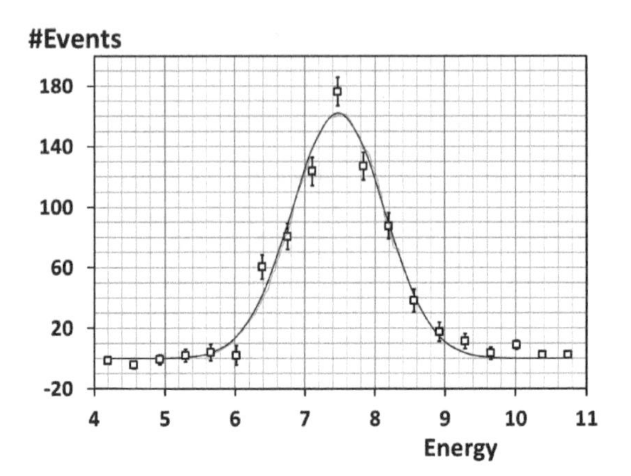

FIGURE 3.17
Distribution of the number of events above background versus neutrino energy based on the data and background in Figure 3.16.

remember that I did not stumble upon this evidence for an 8 MeV neutrino line in the K-II data "out of the blue," but I had been specifically looking for a line at 8 MeV based on this being the condition for the Mont Blanc neutrino burst to be tachyons.

Physics and the Nature of God

A journalist once asked Einstein what he would have thought if the solar eclipse observation testing his theory of general relativity showed the theory to be incorrect. Einstein, had strong opinions on how God designed the universe, including his famous "God does not play dice"[4] In answer to the journalist's question about general relativity being proven wrong by observations, Einstein allegedly replied that "Then I would have to pity the dear Lord. The theory is correct anyway."[5] Here he was probably not being arrogant, but rather expressing the view that his theory was so beautiful that it just had to be right. Regrettably, I cannot say the same, about my 3 + 3 model of the neutrino masses with its tachyonic neutrino. I am not much of a theorist, let alone one who develops beautiful theories. However, I am good at spotting hidden patterns, even though some of the time they may have been only in my imagination.

The contention of this chapter is that the evidence in support of neutrinos being tachyons has been "out there" ever since July 15, 1988, when the K-II group's second article on SN 1987A became publicly available. However, the evidence has been well-hidden due to some exceptionally good camouflage. In fact, it could even be said that evidence for FTL particles has remained hidden ever since Einstein considered them impossible in his 1905 relativity paper over a century ago. Obviously, until Pauli suggested the existence of the neutrino in 1930 no one could imagine they traveled at FTL speed, and this certainly remained true even after neutrinos were detected by Reines and Cowan in 1956. Moreover, given the tiny value of the neutrino mass, m, it has been difficult to determine if its square, m^2, is positive or negative within experimental uncertainties. If some neutrinos are indeed tachyons, it is therefore understandable how the evidence has been undetected for over a century.

Why did I make the idiotic decision to write this book before either a new supernova or the ongoing KATRIN experiment's results either conclusively prove or disprove my 3 + 3 model with its tachyonic neutrinos? I was certainly not relying on the magnificence of some theory that just had to be right. Rather, in addition to the reasons given at the very start of this book, I wrote this book prematurely based on my confidence in the strength of the evidence already found, especially the discovery of a very well-camouflaged 8 MeV neutrino line in the K-II data. In effect, I believed God would have to be extremely mischievous if not outright malicious to have put all those coincidences out there if my model were wrong, and some neutrinos were *not* tachyons. As Einstein has said: "Subtle is the Lord, but malicious he is not."[5]

Summary

Very massive stars end their lives in a massive explosion known as a supernova, in which 99% of the emitted energy is in the form of neutrinos. Supernovae occur only about two or three times a century in our galaxy, and only one (SN 1987A) has occurred since neutrino detectors have become available. The four detectors observing that event in 1987 observed all together only 30 neutrinos. The neutrinos from three of the four detectors were recorded at about the same time, while the fourth one under Mont Blanc mysteriously recorded its burst nearly 5 hours early, and it is usually ignored by most physicists. Discounting the Mont Blanc neutrinos, the data from the bursts in the other three detectors are usually interpreted as placing an upper limit on the mass of the neutrino, assuming the conventional model in which all three neutrinos have nearly the same mass. However, by assuming the neutrinos were emitted nearly simultaneously (that is, during a few seconds), and the spread in their arrival times is mainly due to varying travel times from the supernova, we can calculate the masses of individual neutrinos. The result is two distinct neutrino masses, one around 4 eV and the other around 21 eV. This strange observation was the basis of my unconventional 3 + 3 model proposed in 2013. This model postulated three active-sterile neutrino pairs, one of which is a tachyon with negative m^2. Such an exotic model requires confirmation before it can be taken seriously. The first verification involved finding good fits to the observed dark matter distribution in the Milky Way and in clusters of galaxies using the 4 eV and 21 eV masses.

Further verification for the model comes from the mysterious SN 1987A neutrino burst captured by the Mont Blanc detector. Given its early arrival time and the energy of the neutrinos, we can estimate their mass, and the value is consistent with that of the tachyon in the 3 + 3 model, within a factor of two. But, if these Mont Blanc neutrinos really do all correspond to a single well-defined mass, they would need to have a very precisely defined energy (8 MeV). That is, they would constitute a *line* in the neutrino spectrum, that is, be monochromatic, otherwise they would not reach Earth in a very brief 7 second burst. I initially dismissed this possibility of an 8 MeV neutrino line as being inconceivable when the 3 + 3 model was first proposed. But later I came up with exactly the needed model involving dark matter X-particles that would account for such an 8 MeV neutrino line based on a Z′-mediated reaction. In addition, evidence supporting the model was found in the spectrum of gamma rays from the galactic center. A final key piece of evidence for an 8 MeV neutrino line from SN 1987A was discovered by analyzing 997 events taken by the Kamiokande II detector recorded on the day of SN 1987A. Using other data taken by that same detector recorded many months later to define the background spectrum on the day of SN 1987A, it becomes possible to extract a well-camouflaged 8 MeV neutrino line sitting atop the background – or a "well-hidden unicorn."

The Three Unicorns

The 8 MeV neutrino line hidden atop the background is the first of our three unicorns associated with SN 1987A. It had excellent camouflage because the energy of the neutrino line was very close to the peak of the background. This first unicorn led immediately to its hidden mate – the neutrinos in the 5-hour early Mont Blanc burst being tachyons, which could only be true if that 8 MeV line existed. If you are wondering about the third hidden unicorn, you have already met her, but she is the shiest of all three (note her closed eyes in Figure 3.18).

The third unicorn is the dark matter X-particle of mass 8 MeV. As you recall, the annihilation of these particles in a Z′-mediated reaction, was how we explained the existence of 8 MeV monochromatic neutrinos, that is, a spectral line from SN 1987A. Whether the three unicorns are mythical creatures, or real and just well-hidden, should be resolved by either the next supernova occurring in our galaxy or by a direct experimental measurement of the three neutrino masses. We consider the latter topic in Chapter 5, where we shall attempt to learn which of the following is true: (a) I have been seriously deluded with my 3 + 3 model, (b) God is indeed mischievous, or (c) the model with its tachyonic neutrino is correct. By the end of that chapter, we also shall meet a herd of elephants in connection with the KATRIN experiment.

FIGURE 3.18
The three unicorns associated with SN 1987A.

Source: Image from Shutterstock

References

1. Sagan, Carl. *Cosmos,* New York, NY, Random House, 1980.
2. Kelvin, Lord. *Baltimore Lectures on Molecular Dynamics and the Wave Theory of Light,* London, C.J. Clay and Sons, p. 274, 1904.
3. Ehrlich, Robert. *Tachyonic Neutrinos and the Neutrino Masses, Astroparticle Physics,* 41, 1–6, 2013.
4. From a letter in German to Max Born, published in 1971, Irene Born (translator), *The Born-Einstein Letters,* Walker and Company, New York, 1926: [*Jedenfalls bin ich überzeugt, daß der nicht würfelt.*]
5. Pais, Abraham. *Subtle is the Lord: The Science and the Life of Albert Einstein, Oxford,* Oxford University Press, p. 30, 1982.

4

Theories of Everything and Anything

Warp [superluminal] speed is unfeasible based on absolutely everything we know about the laws of physics. I am 99.99% confident of that.[1]

Sean Carroll, Caltech physicist

Tachyons? "Sure, Why Not"

This book so far has focused largely on the hunt for the tachyon using published data from various sources in particle physics and astrophysics. Specifically, we have considered *empirical* evidence for some neutrinos being tachyons. Until now there has been little discussion of *theoretical* work by those who have tried to show that faster than light (FTL) speed particles exist or argue against the possibility. The latter opinion has been the more common one among theorists, and some have expressed great certainty about the non-existence of FTL tachyons – see opening quote by Sean Carroll. Of course, the truth about our universe is not decided by a majority vote, and other very respected physicists have expressed a belief that superluminal tachyons might exist. A colleague of mine was a graduate student at Caltech during the early 1980s, and he once asked Richard Feynman (see Figure 4.1) whether he thought tachyons existed, to which Feynman responded: "Sure, why not?" Feynman's openness to the possibility proves nothing about the likelihood of their existence, but it might raise questions about assertions of 99.99% confidence in the non-existence of tachyons by theorists – even highly respected ones from Caltech. Considering how many times we have been surprised by the unexpected properties of neutrinos in the past, it could be presumptuous to rule out further big surprises, including their superluminality – especially since no ***sub***luminal (slower than light) neutrinos have ever been definitively observed.

DOI: 10.1201/9781003152965-4

FIGURE 4.1
Photo of Richard Feynman, taken in 1984.

Source: Photo by Tamiko Thiel

The Standard Model of Particle Physics

The standard model of particle physics represents a unification of all the known forces of nature besides gravity. The goal of unification of seemingly unrelated phenomena has a long history, starting perhaps with Isaac Newton who discovered in the 17th century how the same force of gravity that describes falling objects on Earth could explain the motions of the heavenly bodies. Two centuries later James Clerk Maxwell discovered how electricity, magnetism, and light were aspects of a single phenomenon: electromagnetism. A third major step was achieved when Albert Einstein, in his Theory of Special Relativity, unified mass and energy as well as space and time. Now physicists are seeking to find a theory that unifies the four fundamental forces. These four are understood to be the strong, weak, electromagnetic, and gravitational forces. The strong and weak forces are not observed in everyday life, because they only operate at distances comparable to the diameter of the atomic nucleus (around 10^{-15} meters). In contrast, the electromagnetic and gravitational forces have an infinite range. An infinite range means that, for example, the gravitational attraction between you and the Earth decreases the greater your distance from the center of the Earth, but it would never vanish no matter how far away you were from Earth. Strangely, even though gravity is by far the weakest of the fundamental forces, it is the dominant and most obvious one, both in everyday life and the universe at large. I vividly recall a lecture I attended by Feynman at a physics meeting in Washington DC in which he had just finished emphasizing just how weak the force of gravity was when a loud crash was heard when one of the speakers fell off the wall. I never did learn if this was a coincidence, a stunt, or an example of Feynman's "magic powers."

FEYNMAN THE MAGICIAN

"There are two kinds of geniuses, the 'ordinary' and the 'magicians.' An ordinary genius is a fellow that you and I would be just as good as, if we were only many times better. There is no mystery as to how his mind works. Once we understand what they have done, we feel certain that we, too, could have done it. … It is different with the magicians. Even after we understand what they have done, the process by which they have done it is completely dark. … Richard Feynman is a magician of the highest caliber."[2]

The standard model of particle physics has been a major milestone on the road toward unification. Developed in stages starting in the early 1970s, the standard model is a quantum theory of matter that describes the interactions and properties of the 17 fundamental particles shown by names and symbols in Figure 4.2. The three neutrinos, designated in the figure by the Greek letter ν, pronounced "nu," appear in the lower right quadrant of the outer ring. The six particles in that ring can be grouped into two categories: quarks (top half of ring) and leptons (bottom half). The difference between them is that while both quarks and leptons both feel the weak force, only the quarks feel the strong force. The aptly named strong force is so strong that individual quarks can never be pried away from one another and observed in isolation; they are always tightly bound to other quarks. In fact, the neutron and proton, which used to be considered fundamental particles, are now believed to be composites of three quarks. Another category of particles once thought to be fundamental are the mesons, such as the pion and kaon, which are also known to be composites of quarks and their antiparticles, the antiquarks.

The inner red ring in the figure consists of four particles known as "bosons" also called force carriers, because they transmit the forces between the other particles. The force transmission occurs when bosons are exchanged between a pair of quarks or leptons. It is much like you would experience a repulsive force if your workout partner tossed a heavy "medicine ball" to you, or an attractive force if she grabbed it back and you tried to hold onto it. Of the four force carriers, the gluon transmits the strong force between the quarks, the photon transmits the electromagnetic force between charged particles, and the W and Z particles transmit the weak force between leptons or quarks. The Higgs particle (H) is shown at the center of the diagram, befitting its central role in the standard model. It is sometimes said that the Higgs particle gives all the other particles their mass, but more accurately

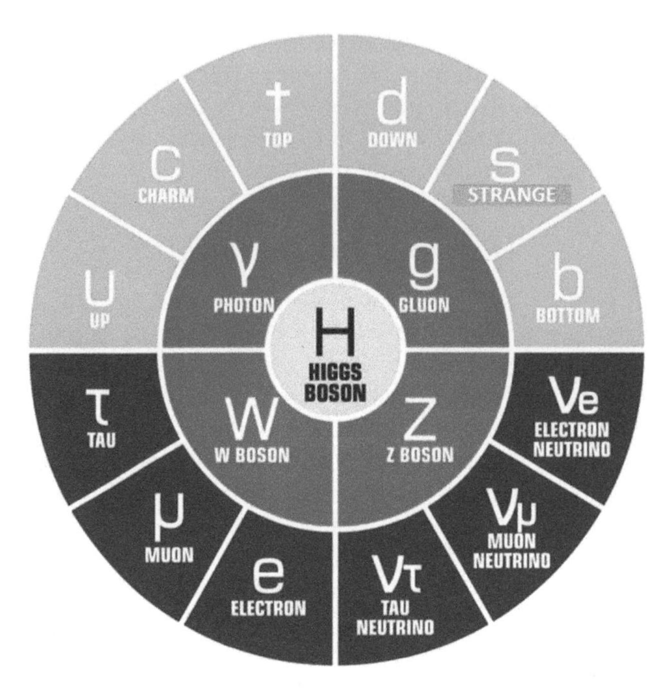

FIGURE 4.2
Standard model of the elementary particles.

Source: Shutterstock image

this particle is the quantum of the Higgs field, much like the photon is the quantum of the electromagnetic field. An omnipresent background Higgs field acts like "molasses" when other particles move through it, and this field, not the Higgs particle itself generates their mass.

The standard model is much more than a simple classification scheme with cute names for particles. The model includes rules for doing calculations that describe just how the particles and the force carriers interact with each other. In fact, if you google the phrase "standard model equation," you will be very surprised what you get. I did not want to include the equation here and scare you! The standard model has been described as highly successful because calculations done using its equation have, apart from a recent possible exception, been found to agree with results from experiment to a very high degree of precision.

A Zoo of Hypothetical Particles and the Uniqueness of Tachyons

All the 17 particles and their antiparticles included in Figure 4.2 have now been observed in experiments, although in the case of the quarks, they have never been observed in isolation, and they never will be according to present theory. At one time, however, all these particles were in the hypothetical category, that is, they arose from a theory or a conjecture before being observed. Sometimes, such theories or guesses were stimulated by some observed anomaly or puzzle, as was originally true for the neutrino long before the standard model was formulated. Despite having been proposed by Pauli in 1930, the neutrino lived for a quarter of a century as only a theoretical suggestion or hypothetical particle before its detection in the 1950s. With the advent of the standard model, hypothetical particles were postulated to fill missing places in Figure 4.2, and they all were eventually found, as was the case with the Higgs particle (2012), the top quark (1995), and the tau neutrino (2000). There are numerous hypothetical particles today, whose existence would require additions to Figure 4.2, that is, extensions to the standard model. A partial list of them would include the: acceleron, axino, axion, bilepton, black hole electron, chameleon, chargino, cosmon, crypton, curvaton, dark photon, dilaton, diphoton, dual graviton, dual photon, guagino, gluino, goldstino, graviphoton, graviscalar, gravitino, graviton, higgsino, holium, inflaton, leptoquark, magnetic monopole, magnetic photon, majoran, micro black hole, neutralino, photino, Planck particle, pleckton, pomeron, preon, pressuron, Q-ball, saxion, sfermion, sgoldstino, skyrmion, sterile neutrino, tachyon, W′ boson, wimpzilla, and Z′ boson.

In some cases, like that of the sterile neutrino or the Z′ boson, there is considerable evidence for the hypothetical particle's existence, but the evidence has not yet reached a threshold for the particle to be regarded as firmly established. For other listed cases, like the diphoton, the particle was originally suggested simply based on an unexpected anomaly found in an experiment, which is now considered to have been a statistical fluctuation. Still other cases include hypothetical particles, like the cosmon, a kind of primeval "super-atom" which contained all the matter in the universe, its existence could not possibly be confirmed.

Among all the current hypothetical particles there is one standout, namely the tachyon. What makes the tachyon unique is that it in fact might be identical to one of the 17 well-established particles, that is, one of the neutrinos. Quite apart from the evidence that one or more neutrinos are tachyons, which was discussed in previous chapters, there is no evidence neutrinos are *not* tachyons. In other words, as was noted earlier, no neutrino has ever been conclusively observed moving slower than light. Similarly, no neutrino has yet been shown to have a positive mass squared. You might naively imagine this unique status of being possibly already included among the 17 in the standard model would make tachyons more acceptable to theorists, but the opposite is true. There are numerous reasons for this skepticism, including the obnoxious theoretical properties of tachyons, and the many mistaken previous claimed sightings. As already noted, theorists have now redefined the very concept of tachyons to be merely instabilities in a quantum field, rather than FTL particles. Some theorists seem to believe that such a redefinition essentially closes the door on the idea of FTL particles, but of course, it does no such thing. You cannot simply rule out FTL tachyons just because you redefine what a tachyon is. Perhaps the disdain for tachyons, that is, the "evil" ones having FTL speed, arises because they were first proposed not from some elegant theory, as with many of the listed hypothetical particles, but instead a simple "loophole" in relativity, as was discussed in Chapter 1.

Beyond the Standard Model

Although the standard model has proven successful in unifying the strong, weak, and electromagnetic forces, there are clear signs of "new physics" beyond it. Three problems with the standard model are: (1) the neutrinos are wrongly predicted to be massless, (2) the nature of dark matter remains unexplained; and of course, (3) gravity is not unified with the other three forces. Additionally, a potential major challenge to the standard model would arise if a new (fifth) fundamental force should be observed. The idea of a fifth force has a long and controversial history going back to the 1980s, when observations suggested a new fundamental force, like gravity, but having a range of about 100 meters. Those observations later proved to be erroneous.

In recent years interest in a possible fifth force has been revived by several cosmological puzzles, especially dark matter and dark energy. Independent evidence for a possible fifth force was provided in 2015 based on a nuclear physics experiment done by Attila Krasznahorkay and his Hungarian colleagues, as discussed in Chapter 3. These researchers reported the existence of a new Z′ boson that has no place in the standard model. Jonathan Feng and collaborators have identified this particle as a "portal" between the worlds of normal matter and dark matter. As was discussed at length in the previous chapter, the Z′ particle played an essential role in the hypothesized Z′-mediated reaction. You may recall that this reaction was the proposed mechanism for generating monochromatic 8 MeV neutrinos from SN 1987A – one of three "hidden unicorns" associated with that supernova.

Another claimed departure from the standard model has been the so-called g – 2 muon anomaly. Muons, essentially ultra-heavy electrons, have a so-called magnetic moment, which causes them to wobble when placed in a magnetic field much like a spinning top. The standard model can predict the frequency of these wobbles for any given magnetic field strength to an astounding eleven-digit precision. Experiments looking for departures from the predicted g – 2 value began at CERN in 1959 at the initiative of Leon Lederman, one of the Nobel laureates who discovered the second type of neutrino. By 2021, following a long series of experiments, the anomaly has almost reached the five-sigma threshold deemed necessary for a major discovery. This failure of the standard model, if genuine, would require the existence of new physics beyond the standard model, and probably a fifth force.

What Monsters Might Be Lurking There?

There have been any number of purported discoveries in particle physics initially described in very dramatic terms, only to later turn out to have been mirages. Whether the g - 2 muon anomaly is hiding a "lurking monster" as suggested by Fermilab physicist Chris Polly or something much more mundane remains to be seen. Although the anomaly had been present for many years, it received a great deal of media attention worldwide during 2021. Such a level of attention would certainly be merited if the anomaly truly requires a new force of nature beyond the known four, even if the anomaly has not yet reached the five-sigma level needed for a discovery. There was in this case high confidence that the experiment was done correctly, since it has been repeated by various groups over many years. However, the theoretical prediction turns out to be ambiguous because there are two alternative ways of calculating the muon value of g – 2 using the standard model, one of which agrees quite well with the experiment. Only by focusing on the alternative calculation does one find any disagreement. This ambiguity was not mentioned in most media accounts of this purported breakthrough, so the anomaly could easily turn out to be a mirage. Sometimes particle physicists may be too eager to bring exciting discoveries to public attention before the case for them is absolutely established. For that reason, even when results satisfy the five-sigma level, and the chances of a statistical fluctuation would be only one in 3.5 million, the purported discovery sometimes turns out to be a mistake.

Replacing Particles by Vibrating Strings and Membranes

The next step beyond the standard model has often misleadingly been called a theory of everything (TOE). Initially, the phrase *theory of everything* originated as an ironic reference to various overgeneralized theories, and it was first used in one of Stanisław Lem's science fiction stories. Later the phrase stuck in popularizations of theoretical physics. As it is used today, a TOE would unify gravity with the other three fundamental forces, making them all manifestations of a single force. Of course, if there is a fifth force, the TOE would need to find a way to incorporate that as well. Among the seven or eight candidates for a TOE the most favored one is probably string theory.

Two theoretical physicists who have been especially active in seeking a TOE are Albert Einstein and Edward Witten. Witten has been a theorist at The Institute for Advanced Study in Princeton, where Einstein had worked a half-century earlier. In fact, Einstein was a founding member of the Institute, when he arrived in the U.S. after having just fled Hitler's Germany. Some researchers consider Witten to be "today's Einstein." He has been the recipient of many awards and honors in both physics and pure mathematics, and he was named "the smartest living physicist" in a 1990 informal poll of cosmologists. Witten is also the developer of M-theory, which is an offshoot of string theory.

In string theory, what we normally consider point particles like electrons are instead assumed to be tiny oscillating strings, or alternatively vibrating membranes. You might be tempted to inquire about the material composition of the strings or membranes, or how they are kept under tension, but these questions have no more meaning than asking what an electron is made of. The strings are not "made of" anything; they are simply the fundamental constituents of matter. In the theory, the various vibrational modes of the strings give rise to the different leptons and quarks. The strings or membranes in string theory are unimaginably small, with a size 10^{-33} cm, making them about a billion trillion times smaller than a proton.

Five consistent versions of string theory have been developed, and M-theory is based on a unification of all five. One weird aspect of string or M-theory is that it requires six or seven extra dimensions – making for a total of ten or eleven in addition to the four dimensions of spacetime. Supposedly, if the universe had more than eleven dimensions it would become unstable and it would collapse into one with only eleven. The meaning of the letter M in the name of the theory is unspecified, and according to Witten, it could stand for "magic," "mystery," or "membrane." On the other hand, since a fundamental version of M-theory is unknown, some skeptics have offered "muddled," as a more appropriate descriptor.

String theory arose in the 1980s, and it initially predicted the existence of FTL tachyons, which theorists sought to eliminate, as being unphysical. To their great relief, theorists found a way to rid their theory of "evil" FTL tachyons through the concept of supersymmetry, also known as SUSY. The combination of string theory and SUSY is known as *super*string theory. According to SUSY, every particle has a counterpart. The names give to these counterparts are obtained by appending the letter "s" to the beginning of the particle's name. Thus, for example, the supersymmetric counterparts of particles, electrons, and quarks are "sparticles," "selectrons," and (my favorite) "squarks." SUSY eliminated tachyons because the theory requires that energy can never be negative, which tachyons would allow. SUSY not only eliminated tachyons, but it also promised to (a) explain the weakness of gravity (a trillion trillion trillion times weaker than the weak force), (b) unify the strengths of the fundamental forces at very high energy, and (c) explain the nature of dark matter. With all these *anticipated* advantages, SUSY has proven to be irresistibly attractive to many theorists, even if it doubled the number of fundamental particles, and it has failed to fully realize (a), (b), and (c).

In the simplest version of SUSY, the energy required to create sparticles is achievable with the Large Hadron Collider (LHC), the world's highest energy accelerator. To the dismay of SUSY supporters, however, no sleptons, squarks, or other sparticles have ever been observed in sexperiments. Of course, theorists can always claim the sparticles have masses too high to have been created in the LHC, so their non-observation does not kill SUSY. However, in the words of one theorist, "SUSY starts getting ugly" for very high sparticle masses. Even worse, none of the extra spacetime dimensions predicted in superstring theory has been observed either, a failure explained by suggesting they are "compactified," that is, they are curled up on such a small scale that we would not detect them. Unfortunately, however, there are many possible compactification schemes, and each one leads to a universe having different physical constants and laws. The exact number of compactification schemes is unknown, but one estimate makes our actual universe just one of about 10^{500} possible ones, comprising in total an incredibly vast *multiverse.*

Astronomer Peter Behroozi and his University of Arizona colleagues in 2019 used a supercomputer to simulate 8 million possible universes, and they found none resembling ours. Eight million may sound like an impressive number, but given a total 10^{500} universes making up the multiverse, that would leave theorists with the task of examining the remaining 99.9999999999999999999 9999999999999999999999999999999999999999999999999999999999999999999999 9999999999999999999999999999999999999999999999999999999999999999999999 9999999999999999999999999999999999999999999999999999999999999999999999 9999999999999999999999999999999999999999999999999999999999999999999999 9999999999999999999999999999999999999999999999999999999999999999999999 9999999999999999999999999999999999999999999999999999999999999999999999 9999999999999999999999999999999999999999999999999999999999999999999999 99999999999999999999999999% of the possible universes. Clearly, it is a fool's errand to sort through the vast multiverse at random to find one resembling our actual universe, no matter how powerful our supercomputers become.

Based on the preceding discussion, M theory is clearly a work in progress, with its multiverse, 6 or 7 unobserved extra dimensions, 17 unobserved sparticles having unknown masses, and no final form. It also has also made zero testable predictions after three or four decades of trying. One might imagine that in reaction to this unimpressive history, many of the world's (estimated 5000) string theorists might have become disillusioned and seek out more promising opportunities. Instead, some of them suggest that we abandon the "old-fashioned" idea that scientific theories need give definite predictions, and just accept their TOE as the best that is achievable. To critics of string theory, most prominently Peter Woit of Columbia University, the enterprise seems more of a theory of *anything* than everything, and since it can never be disproven, it is theology rather than science. The cartoon in Figure 4.3 would seem to be an appropriate commentary on the situation now facing string theorists.

Loop Quantum Gravity

One of the chief competitors to string or M-theory is known as loop quantum gravity (LQG) developed by theorists Carlo Rovelli and Lee Smolin. The theory attempts to develop a quantum theory of gravity using Einstein's

FIGURE 4.3
String theory cartoon.

Source: Courtesy of xkcd comics (xkcd.com)

geometric formulation, but it further assumes space and time are both quantized, that is discrete, rather than continuous. One implication of a quantized space and time is that a minimum distance and time interval exist. LQG also postulates that the structure of space is composed of finite loops woven into an extremely fine fabric or network. The evolution of these networks has a scale on the order of a Planck length, which is approximately 10^{-35} m. LQG differs from string theory in that it is formulated in only 4 dimensions and without the need for SUSY, which are two points in its favor. However, proponents of string theory note that their theory reproduces general relativity and quantum field theory in the appropriate limits, which LQG been unable to do. In addition, they note that LQG has nothing to say about the nature of matter in the universe because its scope is limited to the quantum aspects of the gravitational interaction, and it does not yet include the other three forces. Thus, LQG is also a work in progress, and has no specific experimental predictions. To my knowledge LQG has no more use for FTL tachyons than does string theory, although one can find articles discussing "loop quantum cosmology" (a variation of LQG) and tachyons.

Gerald Feinberg: An Accidental Futurist

Gerald Feinberg was one of the first theorists to work on tachyons, and he developed a quantum field theory for them. He is also credited with naming these hypothetical particles tachyons from the Greek "tachys," meaning rapid. Feinberg was my favorite professor when I was in graduate school at Columbia University, because of the clarity of his explanations, and his willingness to answer questions, which was less common when I was in graduate school than now. Feinberg was also open to highly unconventional ideas both within and outside of physics.

For example, in one of his books "The Prometheus Project,"[3] Feinberg focused on the highly idealistic concept that humanity needed to find a small set of universal goals, and then cooperate to achieve them. He also championed the notion of cryonics, which is the freezing people immediately upon their death, in anticipation that someday their brains could be reactivated. Some cryonics companies were formed in the 1960s and 1970s to offer such services, and at least one still exists. Cryonics may be nonsense, or at least far from currently feasible, but the possibility of reanimating brains may not be quite so far-fetched as it once seemed. In fact, human brain activity has been observed for more than 10 minutes after people have been pronounced dead, according to one study. While no one has tried to reactivate human brains of the deceased, perhaps for ethical reasons, some experiments done with pigs have shown that it is possible to restore some brain function after they have been dead for 4 hours. Cryonics might even prove more feasible than humanity agreeing on a small set of universal goals!

In physics, aside from his tachyon work, Feinberg also predicted the existence of a second type of neutrino (the muon neutrino) that was later observed in the two-neutrino experiment. I was amused to learn that Feinberg, like me, admitted that he developed a serious interest in FTL particles after reading a science fiction tale about them ("Beep" by James Blish). I also recently learned that not only did Feinberg believe that tachyons exist, but also that they might explain the alleged psychic ability of precognition, which he regarded as distinctly plausible. Might he have also gotten a message from the future?

Oliver Heaviside: An Accidental Time Traveler

Although hypothetical imaginary mass tachyons were not suggested until 1962, long before that time, and even well before Einstein's relativity, some physicists were contemplating the possible existence of FTL motion. An early writer about FTL particles is the German physicist Arnold Sommerfeld who in 1904, a year before relativity, named them "meta-particles." However, the earliest person who considered FTL speed particles was self-taught Oliver Heaviside, an English engineer, mathematician, and physicist. Heaviside, in 1888 calculated what would happen if a charged particle moved FTL. Writing 17 years prior to relativity, he found that the moving particle would create radiation in the form of a conical shock wave, much like the wake of a boat traveling faster than the speed of a water wave – thus anticipating the phenomenon of Cherenkov radiation, fully a half century before Cherenkov. In a response to critics who had expressed doubts about the possibility of FTL charged particles because they would require infinite energy Heaviside replied: "Don't be afraid of infinity!" The critics, in fact, were wrong in believing that *classical* mechanics would require infinite energy for an electron to move at FTL speed, something that Einstein deduced later in his theory of special relativity. Heaviside was also the first person to discuss the possibility of gravitational waves in 1893. Such waves were later shown to be a consequence of Einstein's general relativity, but they were not observed until more than a century after Heaviside first raised the possibility.

Aside from his work on FTL speeds, Heaviside contributed to the development of many concepts that we now attribute to others. Thus, he inspired his friend George FitzGerald to suggest what now is known as the Lorentz FitzGerald contraction, and he was the first to publish a derivation of the magnetic force on a moving charged particle, which is now called the Lorentz force. Heaviside also greatly simplified Maxwell's Equations, which originally consisted of 20 equations and 20 variables, expressing them in the modern four equations that we use today.

Since Heaviside is also noted for introducing imaginary numbers in connection with the phase of a wave, George Sudarshan and his "meta relativity" colleagues were lucky perhaps that Heaviside did not realize 50 years before them that particles could have an FTL speed if they had an imaginary mass. Heaviside's biographer Paul Nahin has referred to him as an accidental time traveler in view of his foreseeing so many future advances in physics.

Pavel Cherenkov: An Accidental Nobel Laureate

As already noted, Heaviside in his study of FTL motion also predicted the effect now named after Pavel Cherenkov, who observed it a half-century later. The Cherenkov Effect describes the conical shock wave of light created when charged particles move faster than light in a transparent medium such as water. You may recall from Chapter 3 that effect is how neutrinos are observed in the Super-Kamiokande detector. Of course, only a tachyon could outrace a photon in a vacuum. Therefore, some researchers have looked for tachyons by seeing if any particles emit Cherenkov radiation when traveling in vacuum.

Normally, when an effect is named after a person, as with the Cherenkov Effect, it either was first observed by the person, first explained by them, or investigated in experiments of their design. Remarkably, none of these things is true for this effect. Cherenkov observed the effect that now bears his name in 1934, while working under the supervision of Sergei Vavilov. He was not, however, the first, or even the second person to observe it. As early as 1910, Marie Curie had noticed that radium salts dissolved in water produced a bluish glow, but she did not pursue this observation further. During the late 1920s a French scientist, L. I. Mallet, examined the spectrum of the bluish radiation, and found that it was continuous, rather than consisting of lines or bands that are usually associated with fluorescence. In 1934, Cherenkov's supervisor Vavilov designed some experiments for him to carry out, in the hope of determining the nature of the observed luminescence. Under Vavilov's guidance, Cherenkov concluded that the radiation was the result of light emitted by charged particles moving at a speed faster than that of light in the water. In Russia, Vavilov is usually given greater credit for this discovery rather than Cherenkov. Yet in 1958 Cherenkov shared the Nobel Prize in physics with Illya Mikhailovich Frank and Igor Yevgenyevich Tamm for "the discovery and interpretation of the Cherenkov effect." It is also noteworthy that in his Nobel speech, Cherenkov, while he did mention the work of Curie, Mallet and Sommerfeld and others, neglected to mention Heaviside's work predicting the effect a half century earlier.

FTL Observers and Warp-Drive Spaceships

One provocative question raised about tachyons is whether their imaginary mass is an intrinsic feature, and whether particles traveling faster than light really are any different from other particles. Thus, theorists Erasmo Recami and Roberto Mignani have shown that if it were possible for you to chase an FTL particle with your speed also being FTL, the equations of relativity weirdly require the particle to move *slower* than light.

The idea of FTL observers may sound even more fantastic than FTL particles, because it raises the problem of an infinite energy being needed to reach light speed, which was Einstein's original objection to such speeds. Nevertheless, there is a possible way that such "warp speed" could be achieved based on a solution of the equations of general relativity known as the "Alcubierre Warp Drive," proposed by Mexican physicist Miguel Alcubierre. The key to his idea is that relativity does not impose a speed limit on the expansion of space itself, but only on the motion of an object *within* space. Alcubierre proposed that space surrounding a spaceship might be drastically distorted by a gravitational "warp field" – see Figure 4.4. The ship would essentially be inside a "bubble" or a region of flat (undistorted) spacetime, and the bubble

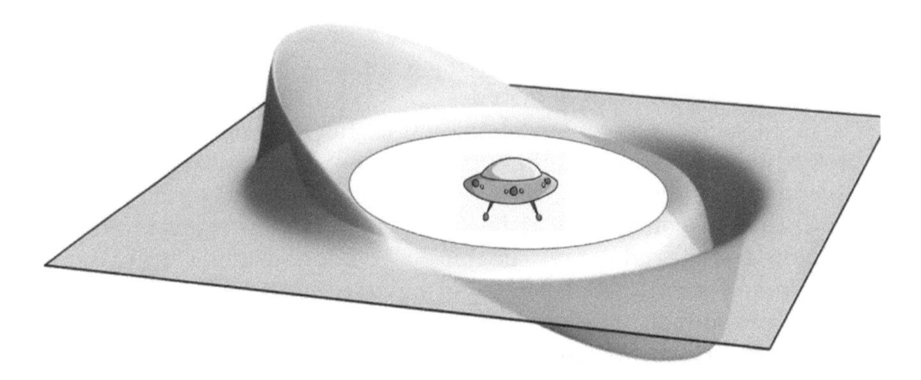

FIGURE 4.4
Warp field according to the Alcubierre drive. Flying saucer added by author, CC-SA 2.0 Germany license: https://commons.wikimedia.org/wiki/File:Alcubierre.png

could then move forward with the ship inside it much like a surfboard rider riding a wave.

For example, if you wanted to travel to a star 1000 light-years away, here is all you need to do according to Alcubierre. First, surround your starship with a warp bubble keeping the space inside flat, and then warp the space in front of and behind the bubble. In a ridiculously extreme case, the warping might shrink the space in front of the bubble from 1000 light-years down to a couple of inches, and then expand the space behind it to 1000 light-years. In this case, the bubble could reach the star after it moved forward just a few inches with the starship at rest inside it. Finally, just pop the starship out of the warp bubble once it reaches the star in practically no time at all. What could be easier?

Among the many difficulties with Alcubierre's idea one big one is that the distortion of space-time he describes would require matter having a negative energy density, which amounts to antigravity. Alcubierre's concept sounds like science fiction in the absence of such exotic matter. However, warp drives have been proposed by astrophysicist Erik Lentz of Göttingen University, which are capable of superluminal motion without any negative energy density matter.

Aside from the hypothetical Alcubierre drive, many other examples of allowed FTL speeds can occur within the framework of relativity. These have been discussed by Erasmo Recami, who has been the most prolific writer of scientific papers and books on FTL speeds, with well over 60 articles, and the earliest known book on tachyons (in 1976). Recami, a retired physics professor from the University of Bergamo (see Figure 4.5), has developed a theory of tachyons and many of his predictions involving superluminal speeds are said to have experimental confirmations.

One of Recami's early papers, written with colleague Roberto Mignani, dispenses with one of the sillier-sounding arguments against the existence

FIGURE 4.5
Pioneering tachyon physicist Erasmo Recami.

of tachyons, namely that since they can have a negative energy, therefore, the universe would be unstable if they existed. Supposedly, in this case vast numbers of positive and negative energy tachyons would be created from nothing, and the whole universe would vanish in an instant, which obviously has not happened, so tachyons cannot exist. Theoretically, such an instability could conceivably occur for so-called "spin zero" tachyons, but it would not occur if neutrinos were tachyons. As we have seen in Chapter 2, according to the "reinterpretation principle," an outgoing negative energy neutrino traveling backward in time is mathematically equivalent to an incoming antineutrino going forward in time. Thus, by that same principle, the emission from some point of a pair of tachyons with positive and negative energy is indistinguishable from a single tachyon simply passing through that point – see Figure 4.6 cartoon. The universe' existence certainly does not imply that

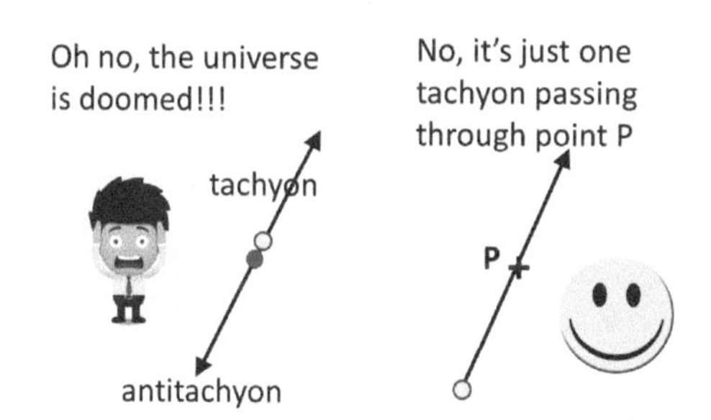

FIGURE 4.6
Why the universe won't vanish even if tachyons exist.

tachyons cannot exist. You can probably sleep soundly knowing that the universe would not disappear if neutrinos should turn out to be tachyons.

Cosmic Inflation

Our observable universe is believed to have begun in a "Big Bang" that occurred about 13.8 billion years ago. Ironically, the Big Bang phrase was originally coined as a term of derision by astronomer Fred Hoyle, an opponent of the theory who was attempting to ridicule it. The first evidence for the Big Bang was the discovery by Edwin Hubble that most galaxies appear to be moving away from us with a speed that is proportional to their distance. Because of that proportionality, if we could play the "universal movie" backward, the galaxies all would converge on a single point 13.8 billion years ago, although more recent data based on the cosmic background radiation have shifted the time since the Big Bang to be "only" 12.5 billion years. The seemingly inconsequential revision was called for because measurements taken from the distant universe were about 10% different from measurements taken from the nearby universe. In many areas of astrophysics, a 10% difference in the value of a constant could be ignored, but here the uncertainties in those measurements were only about 2%. This disagreement in the value of a fundamental constant has according to many cosmologists reached "crisis" proportions, and it has called into question our present picture of how the universe evolved over time.

The best current cosmological theory incorporating the Big Bang has a remarkable feature, known as an "inflationary expansion." Inflation is the term for an extremely rapid expansion of the universe in the first tiny fraction of a second following the Big Bang. During this brief inflation era spacetime itself is said to have expanded trillions of times faster than the speed of light, but no objects *within* spacetime moved at FTL speed. The idea of cosmic inflation, put forth by physicist Alan Guth in 1979, was invented to explain many features of the large-scale universe, including how the seeds for large structures formed from tiny quantum fluctuations.

Because the Big Bang occurred around 12 to 14 billion years ago, we are now receiving radiation from matter that was then located 12 to 14 billion light-years away. Matter at this distance is said to be at our "event horizon," or at the "edge" of our observable universe. Normally, the term event horizon is used in connection with black holes, but here it refers to the sphere surrounding us having a radius at which matter would be moving in all directions away from us at the speed of light. Surprisingly, the two uses of the phrase *event horizon* might be the same, if our observable universe were the interior of a black hole, as some physicists believe. It is conceivable there might be matter beyond our event horizon moving away from us faster than light. In that case, we would never know about it unless it were to slow down, or alternatively if it were emitting tachyons in our direction. In fact, it is also possible that existing black holes in our galaxy might be emitting tachyons, which could exit from within the black hole event horizons.

Dark Energy, Antigravity, and Tachyons

Scientists once thought that the expansion of the universe might gradually slow down over time due to the attractive force of gravity. However, observations show that instead of slowing down, the expansion appears to be speeding up, or accelerating, as though a form of anti-gravity were present. The cause of this accelerated universal expansion is unknown, but it does have a name, namely *dark energy*, which is may possibly be linked to the inflation that occurred in the extremely early universe (during the first 10^{-33} seconds). Figure 4.7 shows a spacetime diagram depicting the evolution of the universe from the Big Bang to the present, which resembles a beautiful crystal goblet. The gentle curvature near the top of the goblet indicates the accelerated expansion occurring now, whereas the rapid increase in the width of the goblet near the stem illustrates the FTL inflation right after the Big Bang. Dark energy, whatever it is, is not a minor constituent of the universe. In fact, based on empirical data, ordinary visible matter comprises roughly 5% the matter/energy in the universe, dark matter comprises roughly 26% and all the rest

FIGURE 4.7
Left: Evolution of the universe from the Big Bang (at the bottom of figure) expanding to the present day on top. Credit: NASA/WMAP Science team. Right: A crystal goblet

(68%) is dark energy. The space-filling stuff known as dark energy might very well be explained in terms of an omnipresent "sea" of tachyons, or specifically huge numbers of tachyonic neutrinos. Here is a simple-minded way to understand the connection between dark energy and tachyons. Tachyons have an imaginary mass, and the gravitational force between any two particles is proportional to the product of their masses. Now, by definition, the product of two imaginary numbers is negative, therefore the forces between tachyons comprising a background sea would be negative, and hence repulsive, driving the accelerated expansion.

The Evidence for Dark Energy and an Accelerated Expansion

The evidence that the universe is expanding at an accelerating rate relies on observations of supernovae seen in other galaxies – see Figure 4.8 which is a plot of the supernova distance versus redshift. The redshift is defined as the increase in wavelength of light due to the motion of the source away from us, and it is a direct measure of the recession speed. If there were no acceleration or deceleration of the expansion, there would be a direct proportionality between the distance and redshift for supernovae, and the data points in the figure would all fall on a straight line. In fact, this is the case for the data

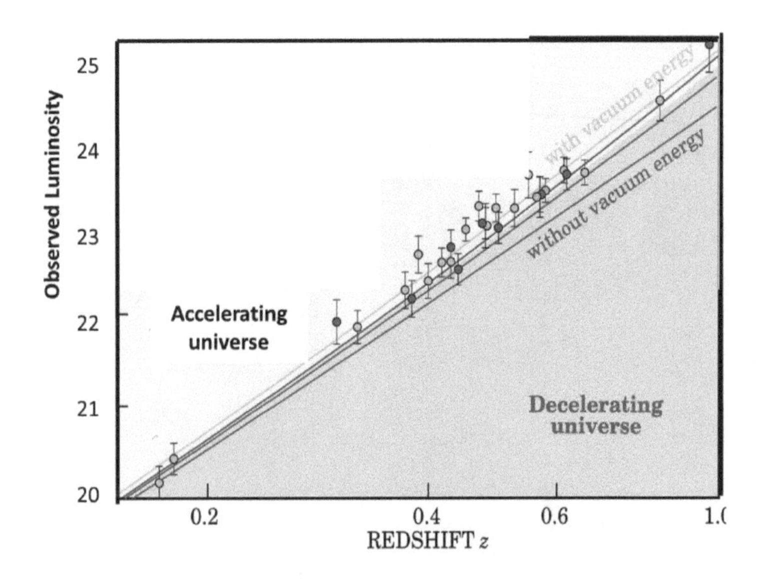

FIGURE 4.8
Graph of observed luminosity (a proxy for distance) versus redshift (a proxy for speed of recession) of supernovae which provided the evidence for an accelerated expansion of the universe based on NASA/WMAP data

points on the left half of the figure, where the redshift is relatively small. Supernovae in the right half of the plot, however, have measured redshifts placing them above the straight line. If the recession speeds are due to the expansion of the universe, then this observation implies that the expansion of the universe is accelerating.

The supernova observations in Figure 4.8 were made in 1998, and they resulted in a Nobel Prize being awarded to Adam Riess, Brian Schmidt, and Saul Perlmutter for their discovery. Subsequently, however, their analysis of the supernova data, and even the very concept of dark energy has been severely challenged. If it should turn out the analysis should be in error, Riess, Schmidt, and Perlmutter need have no worries, because the Nobel Committee does not take back prizes for discoveries later disproven or found inconclusive.

Mirror Universes

One twist on the standard theory of inflationary cosmology was suggested in 2001 by Russian physicists Larissa Rabounski and D. Borissova. According to their theory, at the time of the Big Bang a mirror universe made of antimatter was created and it propagates backward in time while ours advances forward. A spacetime diagram of the evolution of the mirror universe would consist of Figure 4.7 flipped upside down, with the two universes joined at time $t = 0$. Our universe would then evolve forward in time (upward), while the mirror universe would then evolve backward in time (downward). Of course, if there are any beings in the mirror universe, they probably view themselves as the ones going forward in time much as some "down under" Australians consider themselves living on top of the globe.

One advantage of the mirror universe theory is that it answers a question that has long puzzled physicists, namely if matter and antimatter were both created in equal amounts in the Big Bang, where is all the antimatter? On the other hand, some earlier versions of the mirror universe theory do not involve antimatter at all. Those earlier theories put forward in the 1970s suggested that the backward time evolving universe might instead consist of tachyons. All velocities within the mirror universe would be seen by us as being greater than light speed. However, for the inhabitants of that tachyon universe, relativity requires the velocities they observe in their universe would be less than c, and they would judge all velocities within our world to be FTL, if they could observe them at all.

If there really were a mirror universe traveling backward in time it is unclear how its existence could be confirmed regardless of whether it consists of tachyons or antimatter. On the other hand, perhaps one should not underestimate the imagination of some physicists. Peter Gorham and other members of the ANITA Collaboration have been studying extremely high-energy neutrinos from space at the South Pole. In a 2020 paper they reported observing a *single* high-energy neutrino coming up through the Earth. Such an event was considered impossible, given that at its very high energy the neutrino should have been absorbed by the Earth and never reached the detector. Their suggestion, based on this one "impossible event" was that maybe the neutrino was really coming down not up, as it traveled backward in time from the mirror universe!

Other Alternatives to the Standard Cosmology

One other interesting alternative to inflationary cosmology has been proposed by Portuguese-born cosmologist at the Imperial College in London, João Magueijo, who in 1998, together with Andreas Albrecht, claimed that the present universe is not undergoing an accelerated expansion – it just looks that way. They claim what's really happening is that the speed of light is slowing down over cosmic time. Magueijo's varying speed of light theory also applies to the very early universe. He claims that during that early era the universe wasn't expanding much faster than light, rather the speed of light itself was just much larger than it is now – about 10 trillion trillion trillion times larger.

Still another alternative to the standard inflation theory has suggested that our universe began not with a bang but with a "Big Bounce" from a previously contracting cosmos. This Big Bounce cosmology has been proposed by Paul Steinhardt of Princeton University, Martin Bojowald of Pennsylvania State University, and others, and it may be connected to loop quantum gravity. The Big Bounce theory solves one problem with the Big Bang, namely that at the precise moment of the Big Bang there is a singularity of zero volume and infinite energy, or "the end of physics as we know it." However, with the Big Bounce a previously existing universe collapsed, not to a point of singularity, but to some finite size where the quantum effects of gravity become repulsive enough so that the universe rebounds back out. Stephen Hawking also once played with the idea of a collapsing phase of the universe, and its possible connection to a reversal in the direction of time. According to Hawking:

> At first, I believed the disorder would decrease as the universe collapsed … This would have meant that the collapsing phase was like the time reverse of the expanding phase. People in the contracting phase would live their lives backward. They would die before they were born and would get younger as the universe contracted.[4]

Hawking, however, later concluded this very interesting idea was the biggest blunder of his life. Supporters of the Big Bounce theory acknowledge the successes of the inflationary theory in fitting observed data, but they point out that even after fixing all the parameters, any inflationary model gives an infinite diversity of outcomes (a "multiverse") with none preferred over any other, and we happily are in one that permits life to exist. Some cosmologists believe that the vast multitude of universes are all physically real, each in their own bubble immersed in an eternally expanding and energized multiverse. Not surprisingly, the Big Bounce theory is criticized just as harshly by supporters of the standard inflationary cosmology theory. At one time cosmology was justifiably referred to as being metaphysics rather than physics, given the lack of ability to test theories against observations. Now that high-precision measurements are possible that is no longer the case, but clearly cosmological theory is far from settled. The confusing situation for theories of the large-scale evolution of the universe recalls the similarly unsettled state of the theory of everything (or anything) describing the quantum world of very tiny distances discussed at the start of the chapter. It is noteworthy how often the concept of FTL motion comes into play in both instances.

Entanglement: The Effect Einstein Found Spooky

One other example of FTL speed involves the phenomenon known as quantum entanglement. Erwin Schrödinger, one of the pioneers of quantum mechanics who coined the term, has called entanglement "the characteristic trait of quantum mechanics, the one that enforces its entire departure from classical lines of thought." Entanglement occurs when pairs of particles are created in such a way that the quantum state of each of them cannot be described independently from the other, even when they are separated by a large distance. Let us see why entanglement implies FTL speed. Suppose a pair of particles, A and B, created from the decay of a "parent" particle whose spin is known to be zero. In that case, if A has a clockwise spin, that is, or spin "up" about some axis, then B must have an equal counterclockwise spin, that is, spin "down," so that the total spin adds up to be zero. So far this seems straightforward enough, but paradoxical effects can occur if A and B become very widely separated before a measurement is made of the spin of one of them. In such a case, if the two particles are entangled, the quantum state of B becomes known *instantly* as soon as that of A is measured, no matter how great their separation.

Einstein and others considered such behavior very disturbing, and he referred to it as "spooky action at a distance." Einstein, writing in a 1935 paper with colleagues Boris Podolsky and Nathan Rosen (in their famous "EPR" paper), argued that the existence of entanglement meant that quantum mechanics could not be a complete description of reality – a conclusion that many physicists now reject. Einstein and his coauthors noted in their EPR paper that entanglement meant that information about the quantum state of A was being sent from A to B at FTL speed. However, note that the person measuring A does not send any actual signal at FTL speed; rather nature does. For someone to send an FTL message, they would need to control what A's spin was going to be, and in that case, A and B would no longer be a single entangled system.

One way of viewing the preceding example of two entangled particles may make it seem not mysterious at all. For example, you might think that each particle of the pair has a well-defined spin after they separate, and we simply do not know which is the one that is clockwise, and which one is counterclockwise, until one of them is measured, and then the spin of the other instantly becomes known. But this interpretation of each particle having well-defined spin before measurement (a concept known as "realism") is inconsistent with quantum theory, and it has been shown experimentally not to be true. Each particle, until it is measured, must be thought of as existing in a mixed quantum state that is 50% clockwise and 50% counterclockwise, much like Schrodinger's famous cat – or mouse (see Figure 4.9). As you may know, this unfortunate feline is imagined to be in a sealed box containing a sealed vial of poison, which has a 50/50 chance of being broken by a random process. As a result, the cat must be considered in a mixed state that is 50% alive and 50% dead – at least until the box is opened.

FIGURE 4.9
Schrodinger's mouse: "It doesn't matter whether the cat is alive or dead. Just don't open the box!"

Source: Shutterstock image

Quantum entanglement is undoubtedly a real phenomenon, since it has been repeatedly demonstrated in the lab – though of course never with cats, dogs, or anything other than subatomic particles. Entanglement, which establishes a connection between two objects, separated by an arbitrarily large distance, may be directly linked to the existence of wormholes, which also can connect two widely separated spacetime regions – at least so say theoretical physicists Leonard Susskind of Stanford University and Juan Maldacena of the Princeton Institute for Advanced Study. If they are right in their belief that entanglement and wormholes are just two sides of the same coin, this suggests wormholes must be real objects, since we know that entanglement is a real phenomenon. Creating an entanglement of objects much larger than subatomic particles had been thought to pose an immensely difficult technical challenge. However, in 2020, scientists at the National Institute of Standards and Technology (NIST) have amazingly succeeded in entangling two tiny drums. If it were possible to entangle two black holes that were near each other they might form a wormhole. However, it might be necessary that one of the two black holes actually be a hypothetical "white hole" which can be thought of as the reverse of a black hole. White holes are solutions of general relativity which cannot be entered from the outside, but energy and matter can escape from them, but in contrast to black holes, it is unknown if they exist.

Superdeterminism and Bell's Theorem

Ordinary determinism basically says free will is just an illusion. Superdeterminism goes further, and postulates that there is no genuine chance anywhere in the cosmos, and that quantum mechanics with its probabilistic outcomes of observations or experiments cannot be the final word. The "God does not play dice" comment by Einstein was his way of criticizing quantum mechanics as being an incomplete description of nature. In this view, quantum systems are predictable, based on certain as yet unknown "hidden" variables. This alternative, however, seemed to be doomed based on the classic 1964 paper by John Stuart Bell. In his paper Bell proved that quantum physics is incompatible with local (deterministic) hidden-variable theories. Moreover, experiments have shown that quantum mechanics, not hidden variable theories give the correct prediction. On the other hand, the assumption of superdeterminism (SD) would be a loophole in Bell's theorem. According to SD, starting with the Big Bang, everything is causally determined including all our thoughts and actions through some hidden variable theory, yet unknown to us.

The previous section discussed how the phenomenon of entanglement seemed to require that quantum information travels superluminally between

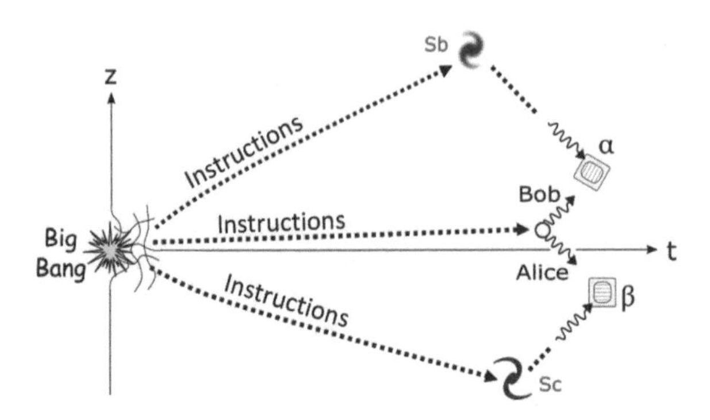

FIGURE 4.10
Illustration of superdeterminism.

two distantly separated points A and B. Thus, when Alice at point A measures the spin of particle 1, this measurement requires that the spin of a second entangled particle 2 be instantly known, even before Bob at point B makes the measurement. Superdeterminism offers an alternative solution. Under SD, however, no information needed to travel superluminally from Alice to Bob, since each of their measurements were independently dictated by prior local influences that could be traced back ultimately to the Big Bang – see Figure 4.10. Since Einstein himself was a determinist who believed free will is simply an illusion, it is interesting that he did not invoke superdeterminism to avoid the need for his "spooky action at a distance." Unfortunately, however, while the concept of superdeterminism may be an interesting hypothesis that could be true, it is not empirically testable, since experimenters would never be able to eliminate correlations between measurements made at two distant locations that were ultimately created at the beginning of the universe. In fact, John Bell himself acknowledged the SD loophole to his theorem, but he also believed that it was implausible. Nevertheless, SD does leave open the possibility that some clever person could construct a hidden variable theory that exactly reproduces the predictions of quantum mechanics.

The Most Famous Failed Experiment in History

Before Einstein's relativity, physicists believed that all of space was filled with some sort of medium, through which light traveled. This medium, which was called the *ether* (or *aether*), would offer no resistance to planets moving through it and would have all the properties of what we now refer to as a vacuum. Such a belief in a space-filling medium seemed reasonable to 19th century physicists, because all known waves, such as sound waves and water waves, required a medium for their propagation, so why not light? Furthermore, it was believed that if we move through the ether, then there should be a kind of "ether wind" blowing against us. Such an ether wind would presumably affect the speed with which light travels, so that the speed would be different when observed in different directions. In fact, such behavior is exactly what we would observe for the changed speed of sound waves in air when a wind blew or when we move through stationary air. Thus, with sound you would only measure the same speed of sound in all directions if there were no wind, and the same was thought to be true for light.

In 1888, a pair of physicists, Albert Michelson and Edward Morley, conducted an experiment to see how the Earth's motion through the presumed space-filling ether affected the speed of light. Essentially, they observed light beams sent from a source along two perpendicular directions by a half-silvered mirror. The light beams were then recombined and viewed through the telescope (see Figure 4.11). Michelson and Morley expected that the motion of the Earth through space would cause an ether wind which would affect the two light beams differently. As a result, it was expected that when the whole apparatus was rotated the recombined light beams should change in appearance. What they observed caused a great shock; there was no effect whatsoever. It was just as if the Earth were at rest, with no ether wind blowing. This most famous "failed" experiment in history was very perplexing, and many scientists tried to explain it. Various ideas were proposed including having the Earth drag some of the ether along with it as it moved (just like it carries its atmosphere along with it), but none of these ideas could explain all observations. Only after Einstein came out with special relativity was the original ether concept finally abandoned. Relativity is based on the assumptions that the speed of light is the same in any frame of reference, and that neither a preferred reference frame nor an ether exists. On the other hand, although it is not well known, Einstein apparently reversed himself

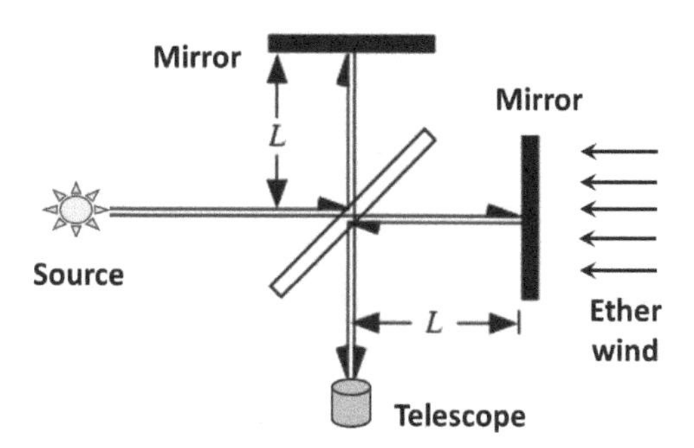

FIGURE 4.11
Basic scheme of the Michelson–Morley experiment in which a beam of light is split by the 45 degree half-silvered mirror, and the two beams are recombined before entering the viewing telescope.

about the existence of an ether after devising general relativity, a decade after special relativity. In a 1919 letter to Lorentz, he wrote:

> It would have been more correct if I had limited myself, in my earlier publications, to emphasizing only the non-existence of an ether velocity, instead of arguing the total non-existence of the ether, for I can see that with the word "ether" we say nothing else than that space has to be viewed as a carrier of physical qualities.[5]

Furthermore, in a 1918 response to critics he wrote: "the 'diseased man' of physics, the 'ether,' is in fact alive and well, but that it is a relativistic ether in that no motion may be ascribed to it." In fact, following Einstein, one could argue that either the hypothetical Higgs ("molasses") field or the observed cosmic microwave background (CMB) serves to define an ether reference frame. The CMB is a relic from the time when the universe was only about 379,000 years old (0.0028% of its present age), which was when it first became transparent to light.

Furthermore, contrary to Einstein's assertions above about not being able to ascribe motion to an ether, we *can* measure our motion through the space-filling CMB because this radiation is 0.12% warmer in the direction of the constellation Leo, and it is 0.12% cooler in the opposite direction toward the constellation Aquarius. This observation tells us that the Earth (averaged over a time of one year) is moving at 0.12% the speed of light in the direction of the constellation Leo through the cosmic background radiation. Apart from the variations due to our motion through the CMB, its temperature is almost (but not quite) uniform in all directions.

It may seem strange that observations of the CMB allow us to determine our absolute speed through space, whereas the negative result of the Michelson–Morley experiment suggests that this is not possible. Recall that in the Michelson–Morley experiment light sent along any pair of perpendicular directions traveled at the same speed, just as though we were at rest in an ether. The difference between this negative result and the CMB observations is that the latter is a direct measurement of our speed, whereas the Michelson–Morley experiment using the travel time of perpendicular light beams depends on a hidden assumption involving synchronization of clocks at different locations. If we try to measure the "one-way" speed of light, from a source to a detector we need to have a convention as to how to synchronize the two clocks located at the source and the detector. In contrast, the Michelson–Morley experiment can be shown to be equivalent to a "two-way" speed of light measurement in which light is sent from the source to the detector, and then back to the source. Such a two-way speed measurement needs only *one* clock, so it makes no assumptions regarding synchronization of clocks at different locations. Albert Einstein in devising relativity recognized this problem, and he simply *chose* a synchronization convention which *assumed* the one-way speed of light equal to the two-way speed. A weird example may make the distinction clearer between one-way and two-way light speed measurements. Suppose it is not possible to measure the one-way speed, which in fact has never been done. In that case, it is theoretically possible that the speed of light is infinite in one direction in space and c/2 in the other direction. In such a case, the round-trip (two-way) speed, computed from the round-trip travel time and distance, would be the same if its speed were identically c in both directions, which is what Einstein assumed to be true.

Einstein's Ether

The preceding discussion explained that Einstein in 1918 explicitly recognized the existence of an ether, noting that: "the 'diseased man' of physics, the 'ether,' is in fact alive and well, but that it is a relativistic ether in that no motion may be ascribed to it." Einstein, however, passed away prior to observations of the CMB, which in fact *does* allow us to ascribe a velocity with respect to a preferred reference frame in space. Starting in the 1980s a so-called Einstein ether theory was developed which includes Einstein's general relativity, but also has a preferred reference frame. These theories also included a universal notion of time, and they violate the principle known as Lorentz Invariance or Lorentz Symmetry, which states that the laws obeyed by any physical system cannot depend on the state of uniform motion. Lorentz invariance also requires that those laws must be written in such a way that they do not depend on quantities that vary from one reference frame to another, such as a particle's energy, speed, or position.

The possibility that Lorentz invariance may be slightly violated is currently unresolved, and it has been actively investigated in recent years by various theorists, including Alan Kostelecký, Stuart Samuel, and Don Colladay. They note that such a possible violation might be expected to show itself in various ways, including superluminal particle speeds. In some models of violated or "broken" Lorentz symmetry, it is postulated that a background field was created shortly after the Big Bang. This background field would cause particles to behave differently depending on their velocity relative to it, effectively defining a preferred reference frame.

Hints of possible violations of Lorentz Symmetry have come from the highest energy neutrinos ever observed in an experiment known as *IceCube* – *see* Figure 4.12. The IceCube Neutrino Observatory, located in Antarctica, has thousands of sensors under the ice, which are distributed over a cubic kilometer (quite an ice cube!) The purity of Antarctic ice makes this location ideal for observing the Cherenkov radiation created from high-energy neutrino interactions in the ice. This experiment was the first to have observed extremely high-energy neutrinos from extragalactic sources. In one analysis of the spectrum of those observed neutrinos, Jiajun Liao and Danny Marfatia have claimed that the spectrum is well explained if a Lorentz violation is present, and that some of the neutrinos are indeed superluminal.

FIGURE 4.12
The IceCube Neutrino Observatory at the South Pole.

Source: Martin Wolf, IceCube/NSF

Making Tachyonic Neutrinos Less Obnoxious

A violation of Lorentz Invariance, in addition to having observable consequences such as FTL neutrinos, also has important implications for theorists who attempt to construct equations describing them. A common approach by such theorists is to modify the equation that Dirac originally used for electrons. Dirac found that his equation had two solutions corresponding to particles with equal and opposite electric charge. On this basis, Dirac predicted the existence of the positively charged electron called the positron, which was later observed. In a similar spirit, theorist Alan Chodos (see Figure 4.13) found that a modified Dirac equation predicts two solutions having equal and opposite sign values of $\pm m^2$, corresponding to bradyons and tachyons having the same magnitude mass – one real and the other imaginary.

Not all theories of tachyons violate Lorentz Invariance and causality, including one developed by Polish theorists Jakub Rembielinski and Jacek Ciborowski, who in 1996 developed a quantum field theory of tachyonic neutrinos. Their theory, however, does assume a preferred frame of reference. It also preserves causality, which means that tachyonic neutrinos could not be used to send messages back to the past. Moreover, in their theory tachyons cannot have negative energy. Marek Radzikowski and separately Ulrich Jentschura have also developed similar quantum field theories for tachyons. With their theories, Chodos, Jentschura, Rembielinski, Ciborowski, and Radzikowski, as well as others, have addressed many of the issues with

FIGURE 4.13
Tachyon hunters Alan Chodos, Marek Radzikowski, and the author at the 17th international conference on tachyons in 2017 planning for the 16th one.

tachyonic neutrinos, making them less obnoxious in terms of their properties. Erasmo Recami, and more recently, Charles Schwartz, a Professor Emeritus of the University of California at Berkeley have also been particularly active in trying to debunk anti-tachyon biases of many physicists. Those who argue that FTL neutrinos are unphysical note that they would violate the basic precepts of physics but as we have noted, this need not be the case. I recently asked Ulrich Jentshura what he considered the strongest remaining theoretical argument against the existence of tachyons, or more precisely, some neutrinos being tachyons, and his answer was there isn't any.

Logical Inconsistency of Experimental Results

Sometimes theoretical arguments can show that an experimental result is impossible, or at least highly suspect. Recall that in 2011, the OPERA collaboration had just startled the world with its claim that they had detected FTL neutrinos – a claim they later retracted. Before that retraction, theorists Andrew Cohen and Sheldon Glashow showed that most of the neutrinos in the OPERA beam would have been removed from the beam had they been tachyons, and they would never have reached the detector. The beam depletion would be the result of Cherenkov radiation, which only FTL particles could emit in vacuum. Their calculation showed that the initial OPERA claim could not be correct. It is important to note, however, that Cohen and Glashow did not rule out tachyonic neutrinos in general by their calculation, just those having the very large imaginary mass that was implied by the OPERA initial result. Theorists were properly skeptical about the initial OPERA claim of FTL neutrinos, especially given the large value of the imaginary mass implied. Nevertheless, one should be cautious when it comes to using theoretical arguments to debunk observations as impossible. As Gerald Feinberg has noted: *particles that travel faster-than-light do not involve logical inconsistencies. Indeed, no observations can be logically inconsistent.*

A well-known physics joke illustrates Feinberg's point. An experimentalist tells a theorist that her experiment has shown that X is true. The theorist, on hearing this, replies that he can explain why X must be true. However, the experimentalist soon discovers she has made a mistake. After realizing this fact, the experimentalist tells the theorist that what she really meant to say is her experiment in fact showed X to be *false*. The theoretician after a moment's thought replies that he has also made a mistake, and that he can prove X must be false. Clearly, experimental results can be mistaken as was the case for OPERA's initial report of FTL neutrinos. On the other hand, sometimes when observations are found to be inconsistent with a current theory, the theory is wrong. In such cases, some clever theorist will usually come up with ways to accommodate whatever experiments have revealed about our wondrous universe. Physics would be doomed if it ever gives up the need to test theories experimentally, as some string theorists have suggested.

Summary

This chapter is about the many ways that FTL particles and superluminal speeds come up both in theories of the very small (superstring theory) and the very large (inflationary cosmology). Some of these ideas are generally accepted, while others are far more speculative. Regarding the small-scale universe, the quest of many theoretical physicists has been the unification of the four fundamental forces into a "theory of everything" or TOE. A major step along the way toward a TOE has been the very successful standard model of particle physics. In the opinion of many theorists the most promising candidate for a TOE has been eleven-dimensional M-theory, which is a unification of the various forms of superstring theory. However, M-theory is not completely defined, and worse, it may not be testable, leading some critics to claim it is more of a theory of anything than everything.

Tachyons initially routinely popped up in string theory. However, they were banished after the idea of supersymmetry or SUSY was incorporated, leading string theory to become superstring theory. The price for this tachyon banishment was, however, to double the number of subatomic particles since each known particle must have a partner – a "sparticle." Unfortunately for SUSY, however, no evidence supports the existence of these sparticles nor for the extra seven dimensions in M-theory. String theory, as currently understood, does have a place for imaginary mass tachyons but these particles do not have superluminal speed and merely represent an instability in the quantum field, such as that associated with the Higgs particle. Loop quantum gravity is another leading candidate besides string theory for a TOE but it also has made no experimentally verifiable predictions.

Various physicists have explored the possibility of FTL particles, even before special relativity, and some researchers have even explored the possibility of FTL travel or "warp speed," which might be conceivable based on general relativity, provided exotic matter exists having negative energy density or negative mass. Another example of FTL speeds involves the expansion of space itself during the very early universe, a phenomenon known as cosmic inflation, which is the prevailing model of how the early universe evolved. Inflation is not forbidden by relativity, because it is an expansion of space itself, not something moving through space.

The force driving the current accelerated expansion of the universe is referred to as dark energy, essentially a form of antigravity. The nature of dark energy is unknown even though it comprises 68% of the total mass/energy in the universe. Dark energy may also be responsible for the sudden accelerated expansion of the early universe – the era of inflation. Tachyons, and specifically a sea of tachyonic neutrinos, might well account for the omnipresent dark energy. Still another example of allowed FTL speed involves the phenomenon of quantum entanglement. This effect, which Einstein found "spooky," involves instantaneous changes in the quantum state of one particle when its companion entangled particle is observed – even if the two are separated by vast distances. However, entanglement, which has been directly observed, does not allow us to send any information or particles at FTL speed.

Lorentz Invariance is the principle that there is no preferred frame of reference, so that the laws of physics are the same in all uniformly moving reference frames. Some theorists think that Lorentz Invariance may be slightly violated in neutrino reactions, and the IceCube experiment may have in fact observed superluminal neutrinos, which would be one sign of such a violation. A slight violation of Lorentz Invariance or alternatively a preferred reference frame appears to be needed in most (but not all) theories of neutrinos as tachyons.

Despite all the theoretical arguments for and against some neutrinos being tachyons, the question ultimately needs to be resolved experimentally. If we were lucky enough to observe a new supernova occurring in our galaxy this might resolve the matter, but such events occur only a few times per century. In terms of Earthly experiments, a precise measurement of either the neutrino velocity or mass also could potentially resolve the matter, with the mass measurement probably being a far more sensitive test than one of speed. However, the opposite would be true for supernova measurements, particularly if the next supernova were to yield an early burst akin to that found in the Mont Blanc detector. Such a burst could no longer be dismissed if it consisted of 1000 neutrinos. In any case, given the rarity of galactic supernovae, the neutrino mass experiments discussed in the next chapter may offer the best hope of a near-time resolution of the central question of this book: Is the tachyon which we have called a unicorn truly a mythical creature, or is it just exceptionally hard to observe?

References

1. Quoted by Gene Weingarten, Washington Post, June 11, 2020.
2. Kac, Mark. *Enigmas of Chance: An Autobiography*, New York, NY, Harper and Row, 1985, p. xxv.
3. Feinberg, Gerald. *The Prometheus Project: Mankind's Search for Long-Range Goals*, New York, NY, Doubleday Anchor, 1969.
4. Hawking, Stephen. *Illustrated Theory of Everything: The Origin and Fate of the Universe*, Beverley Hills, CA, Phoenix Books, p. 95, 2002.
5. Barone, Michael, and Franco Selleri (eds.), *Frontiers of Fundamental Physics*, Berlin/Heidelberg, Germany, Springer Science and Business Media, pp. 193–201, 1993.

5

Weighing the Gravitophobic Neutrinos

Neutrinos have mass? I didn't even know they were Catholic![1]

Author Dan Brown ***in Angels & Demons***

Introduction

Unlike those people who fear stepping on a scale, neutrinos are not literally afraid of being weighed, or gravitophobic but it almost seems that way, given the extreme difficulty in weighing them. As of early 2021, all subatomic particles except for the neutrinos, have well-measured masses. Electrons, for example, have a mass known to an astounding 10 decimal place accuracy. Their mass, 0.5109989500 MeV, is so accurately known that if we applied that accuracy to a passenger jet, we would be able to detect a mosquito as a stowaway just by weighing the plane.

In contrast, physicists were not even certain that neutrinos had a (nonzero) mass until recent decades. Initially, these particles were assumed massless just like photons, as had been specified in the standard model. This assumption became untenable once it was found in 1998 that the various neutrino flavors could oscillate into one another. The idea is that the oscillation frequency for any pair of neutrino flavors is proportional to the differences in the squares of two masses, that is, the quantity $m_1^2 - m_2^2$. Therefore, if all three neutrinos had zero mass, the frequencies would all be zero, so no oscillations.

It does seem strange, however, that the masses of some particles like electrons can be measured with incredible precision, yet for neutrinos their m^2 values are so close to zero that we are not even certain if their masses are real or imaginary, that is, whether their m^2 is positive or negative. The combination of the weakness of the neutrino interaction with other matter and the smallness of their masses is what has made even finding an upper limit for their mass extraordinarily difficult. In addition, neutrinos present the further challenge that they are always weighed without even observing them! What is invariably done is to observe some other particles with which the neutrino has interacted, and then infer the neutrino's mass based on that interaction. One type of observation involves the cosmic microwave

DOI: 10.1201/9781003152965-5

background radiation (CMB), which consists of electromagnetic radiation that fills all space. The CMB, poetically called a "last faint whisper" of the Big Bang, has now cooled to about 2.7 °C above absolute zero, that is –273 °C, but the observed temperature varies very slightly in different directions in space (after we remove the variation due to our motion through the galaxy). Those tiny temperature variations supposedly tell us about the degree of clumping of neutrinos and other particles in the very early universe, which in turn can be related to the sum of masses of the three kinds of neutrinos. Based on the best current CMB data, the sum of the neutrino masses is believed less than about 0.1 eV. However, this upper limit on the sum depends on the correctness of the model of how the early universe evolved over time, and on the ten adjustable parameters that go into the model. Instead of relying on model-dependent results from cosmology, it is preferable to obtain neutrino masses directly from lab experiments which are less model-dependent. The rest of this chapter focuses on these experiments, and how an imaginary ("tachyonic") mass might be revealed. We also consider whether the 3 + 3 model of the neutrino masses with its tachyonic neutrino, one of the three unicorns in Chapter 3, is consistent with the results of neutrino mass experiments. Before the end of this chapter, we will encounter a herd of elephants hidden in the morning mist.

Weighing the Muon Neutrino

Although our main concern will be with the electron neutrino, beginning with the muon flavor is simpler. Recall that the muon neutrino is that second type discovered in the two-neutrino experiment. The muon neutrino, like the other flavors, is a mixture of three neutrinos, but they are normally assumed to be so close in mass (see Figure 1.11) as to be indistinguishable. Experiments to measure the muon neutrino mass invariably study the process depicted in Figure 5.1.

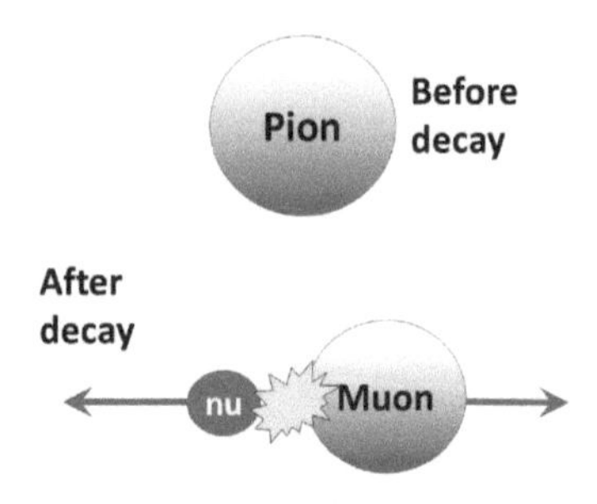

FIGURE 5.1
A stationary pion decays into a 31 MeV lighter muon and a neutrino, each carrying a fixed share of the released energy.

Here a stationary pion decays into a muon and a muon flavor neutrino traveling in opposite directions. Since the pion has been measured to be 31 MeV heavier than the muon, the mass of the muon neutrino determines how much of that 31 MeV mass difference is converted to kinetic energy of the two emitted particles. In one extreme case if the muon neutrino mass were 31 MeV there would be nothing left for kinetic energy, and both the muon and the neutrino would be created at rest when the pion decays. Thus, the very fact the process occurs at all means that the muon neutrino mass is no more than 31 MeV.

To measure the neutrino mass, we need to only find the emitted muon's kinetic energy. We do not observe the neutrino also, because the muon and neutrino recoiling in opposite directions each carry a fixed fraction of the total available energy. A measurement of the muon kinetic energy could be made by simply measuring the length of a track the particle leaves in some medium before coming to rest.

MAKING SUBATOMIC PARTICLE REACTIONS VISIBLE

Bubble chambers are devices in which charged particles leave tracks consisting of trails of tiny bubbles as they move through some liquid. According to its inventor, Donald Glaser, early prototypes of the device were filled with beer, but mostly liquid hydrogen is now used. Since the chamber is usually in a magnetic field charged particles follow curved paths as they are deflected by the field, while neutral particles leave no tracks and are not deflected. Figure 5.2 is an example of a photo of the tracks found in a bubble chamber when the device is placed in the beam from a particle accelerator. An experienced analyst of this picture would easily identify that the following sequence of events occurred. A positively charged pion was created at point A, left a nearly circular track, eventually came to rest at point B, and then decayed into a muon and a muon neutrino. The muon left a very short track before it also came to rest. Finally, the muon decayed creating a positron which created the pretty spiral track. The inward spiral reflects its gradual energy loss. Neutrinos, of course, leave no tracks in the detector, being uncharged particles.

From a careful measurement of the length of the short muon track, we could deduce the muon's original kinetic energy and finally calculate the neutrino mass based on conservation of energy. Suppose we had a device that carried out this process automatically, and then displayed the result as the height of a red column like a thermometer – see Figure 5.3. The figure also shows five possible scenarios for pion decay corresponding to progressively larger kinetic energies of the muon. The top row (case 1) shows the extreme case we previously considered, where the muon is at rest after the pion decay. Here the neutrino must also be at rest, so as not to have an unbalanced momentum. We have already seen that in this case the neutrino mass

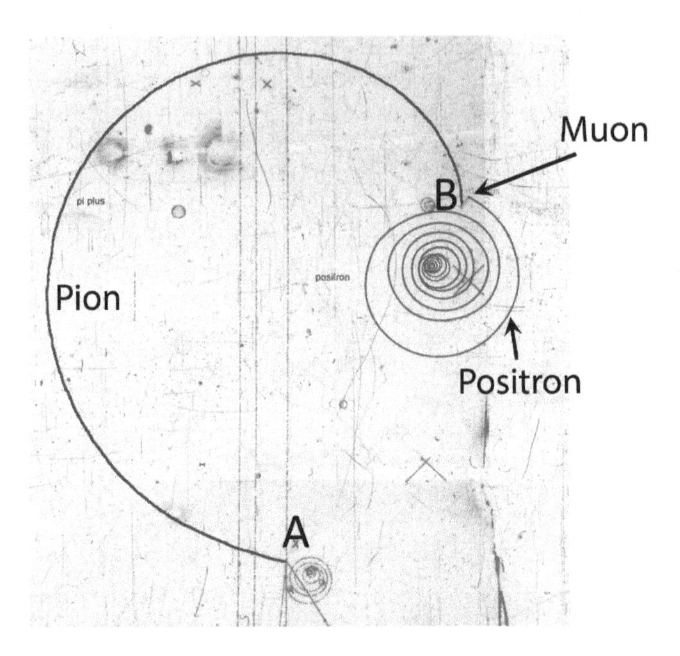

FIGURE 5.2
Photo of pion decay for the CERN 2m hydrogen bubble chamber.

would be its maximum possible value of 31 MeV – which is the "Max" reading on the scale.

In case 2, the muon kinetic energy is a very small value. If this result were found, the mass of the neutrino would have to be a bit less than the maximum 31 MeV since some mass lost has been converted into the kinetic energy

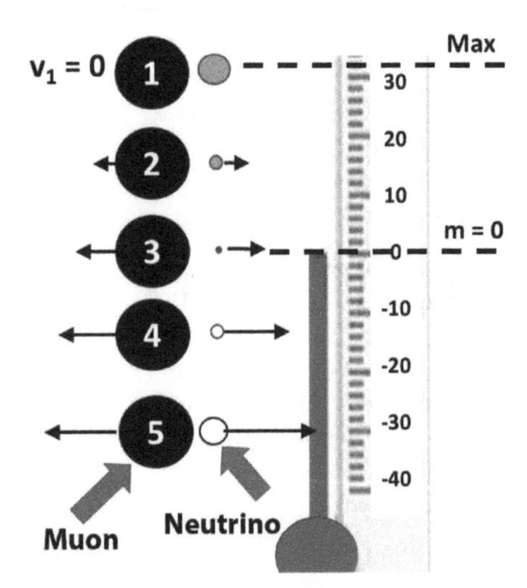

FIGURE 5.3
Pion decay into a muon and a neutrino for five possible neutrino masses.

of the muon and neutrino. For some high enough observed muon energy (case 3) all the 31 MeV mass lost is converted to kinetic energy, leaving nothing left for the neutrino mass. Thus, for case 3 the neutrino would be deduced to have zero mass, and we see a scale reading of zero which happens to coincide with the position of the top of the red column in the figure. What if a muon energy above 31 MeV were observed, as in case 4 or 5? It can be shown by simple algebra that the equations of energy and momentum conservation would then yield an *imaginary* (not negative) mass, and the muon neutrino would have to be a tachyon. In principle, we therefore can learn if the neutrino mass is real or imaginary by measuring only the muon kinetic energy in the decay of a stationary pion. The important lesson here is that the same device can yield real or imaginary values for the muon neutrino mass, which may remove some of the weirdness of the meaning of imaginary mass.

Researchers have performed these pion decay measurements, and as of 2021, the average value for all such experiments is $m^2 = -0.016 \pm 0.023$ MeV^2 for the *square* of the mass of the muon neutrino. The negative value of m^2 here is equal to zero within the quoted uncertainty, so this result certainly does not prove that the muon neutrino is a tachyon. For us to conclude that m^2 were truly negative the value for m^2 would need to be at least five times further from zero than the quoted uncertainty, under the "5-sigma" rule used in physics to claim a discovery. Even if such a result were found, virtually all physicists would want to see it independently confirmed. They would also justifiably want to be certain there were no systematic effects that might account for such a result, given its profound implications. In fact, at one time there were two different possible values for the pion mass, one of which would have implied that the muon neutrino was a tachyon.

A Scale for Measuring Imaginary Mass

It is sometimes suggested that quantities with imaginary values, like the mass of a tachyon, are not directly measurable, but that is not true, as we have just seen. It may be impossible to measure the imaginary mass of a tachyon by literarily putting it at rest on a scale, because tachyons cannot be brought to rest and remain tachyons. However, no subatomic particles ever have their mass found by putting them on a scale – despite the cartoon in Figure 5.4. Rather, as we have seen, the masses of particles are usually found based on their interactions with other particles whose masses are known. In the last section we explained how a hypothetical "scale" might work that could determine the muon neutrino mass based on pion decay, and we saw how the same device would work equally well for neutrino masses that are either real or imaginary. In fact, to add even more realism to our imagined device, the readings might jiggle a bit from one pion decay to the next. That variation

FIGURE 5.4
Cartoon by Geoff Elkins, https://Geoffelkins.wixsite.com/home

occurs because the initial pion will never be exactly at rest, but invariably has some small random motion. This would be analogous to how your observed weight on a scale would jiggle a bit as you shift your position slightly.

In the pion decay experiments to measure the mass of the muon neutrino, each *individual* decay allows us to obtain a value of the mass based on just that one event, but of course, the more events we observe, the more accurate the measurement would be. Unlike the muon neutrino, finding the mass of the electron neutrino cannot be done using individual events. Instead, it is found based on a statistical distribution of *many* events. The difference in the two cases occurs because unlike pion decay, which gives rise to only two particles, the electron neutrino mass is measured in the process of beta decay which results in three particles not two. In beta decay, a radioactive "parent" nucleus decays into a "daughter" nucleus together with an electron, and an antineutrino. In this case, for a decay into three particles each of them no longer gets a fixed share of the released energy from one decay to the next – sometimes one particle gets the lion's share and sometimes another one does. You may recall that the existence of a continuous range of electron energies in beta decay was exactly the mystery which Pauli solved by postulating the neutrino in the first place. He reasoned there must be some unobserved neutral particle (the neutrino) emitted along with the electron that carried off some of the energy. Technically, in beta decays in which an electron is emitted, it is always accompanied by an antineutrino rather than a neutrino, but this distinction will be ignored in most of the rest of this chapter. In fact, it is usually assumed that every particle and its antiparticle have the same mass.

The Beta Decay Spectrum and the Electron Neutrino Mass

The usual way to learn about the mass of the electron neutrino is through its impact on the observed shape of the beta decay spectrum for the emitted electrons. The most common radioisotope used in these experiments is tritium, an isotope of hydrogen containing two neutrons along with one proton, which decays into a helium-3 nucleus, an electron, and an antineutrino, as in Figure 5.5.

Figure 5.6 shows the tritium beta spectrum of the emitted electrons. The figure shows that the most likely electron energy from the decay, that is, the peak of the spectrum, is about 3.5 keV (thousand electron volts). Moreover, the highest emitted electron energy, known as the spectrum endpoint, is E_0 = 18.574 keV (18,574 eV). This energy is also known as the "Q-value" of tritium, which is the kinetic energy released in the decay if the neutrino mass were zero.

The maximum possible energy for the emitted electron, of course, occurs when the emitted neutrino has as little energy as possible, namely just its rest energy, mc^2. Thus, if the neutrino mass happened to be 1, 2, or 3 eV, energy conservation would require that the spectrum for the electrons would end 1, 2, or 3 eV before 18,574 eV. In principle, therefore, we can deduce the neutrino mass by observing where the spectrum is found to end. This idea is illustrated in Figure 5.7 which shows what the tail end of the spectrum would look like, if the electron neutrino mass had three possible values: 0, 1, or 2 eV – see the three labeled solid curves. The horizontal axis of this plot shows values of the electron energy relative to 18,574 eV, so an energy labeled –1 is really 18,573 eV. Obviously, Figure 5.7 is an extreme blow-up of the region within 3 eV of the spectrum endpoint. As the figure shows, the three curves for the masses 0, 1, 2 eV go to zero, that is, have a spectrum endpoint, at energies 0, 1, 2 eV before the point labeled zero. In practice, however, we cannot really deduce the mass by simply seeing where the spectrum goes to zero, because the Q-value for tritium (and hence the energy labeled zero)

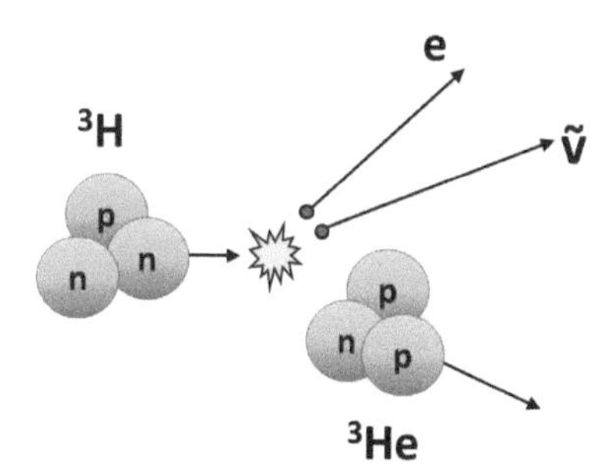

FIGURE 5.5
Tritium beta decay into a helium nucleus, an electron, and an antineutrino.

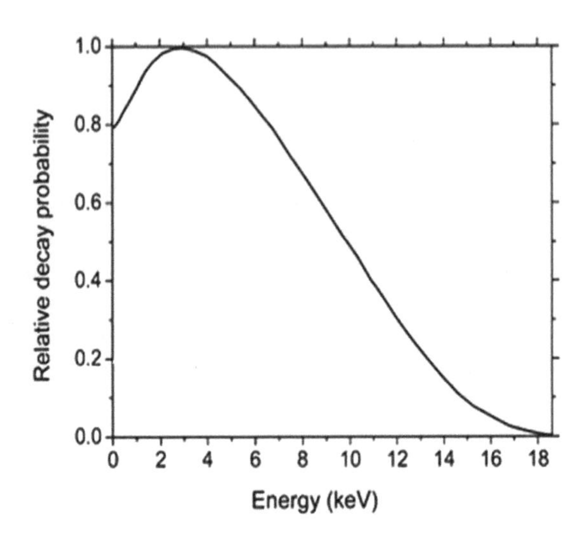

FIGURE 5.6
The energy spectrum of electrons emitted in the beta decay of tritium.

is not exactly known. Instead, the mass is found by fitting the shape of the observed spectrum near its endpoint, and then seeing what values of m and E_0 best fit the data.

One difficulty in carrying out this analysis is that the region near the endpoint contains only a very tiny fraction of all beta decays. For example, the shaded region in Figure 5.7 (the final 1 eV) corresponds to a mere 0.2 trillionths of all decays for the m = 0 spectrum. So, to observe one event in that

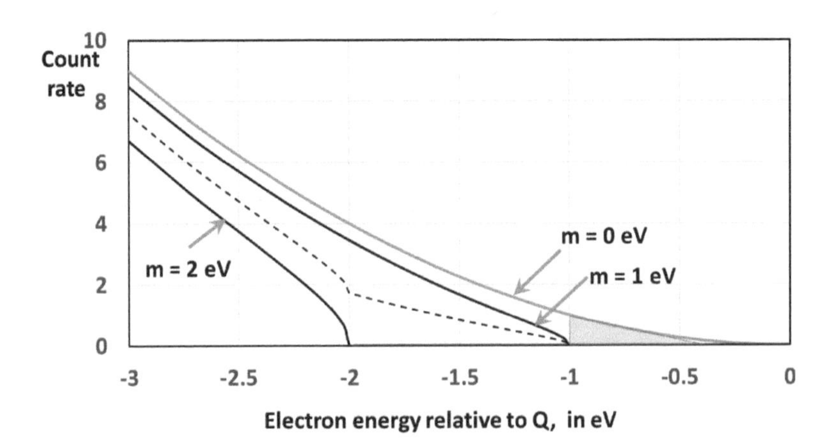

FIGURE 5.7
Shape of spectrum near the endpoint for 3 neutrino masses: 0, 1, and 2 eV. The dashed curve is an average of the m = 1 eV and m = 2 eV curves. Energies are in eV relative to 18,574 eV, and the count rate units are arbitrary.

last 1 eV we would need to have about five trillion beta decays occur – assuming a zero-mass neutrino. Of course, if the mass of the neutrino were 1 eV or more, *zero* events would lie in that last 1 eV, apart from some ever-present background counts. In practice, the presence of background would raise all the curves depicted in Figure 5.7 by some constant level, because the background typically varies very little over the small energy range shown.

Tritium is most often used as the radioisotope in neutrino mass experiments because it combines the two advantages of having a very low Q-value or small spectrum endpoint energy and a not overly long half-life of 12.3 years. These two factors result in more events near the end of the spectrum, and they allow a more precise mass determination. Tritium does, however, have several properties, which make it a challenging one with which to work. For example, among all naturally occurring substances its monetary value (about $30,000 per gram) is comparable to that of (not very high quality) diamonds. In addition, tritium is an important component in nuclear weapons. As a result, for safety, economic, and security reasons, experimentation with tritium demands an entirely closed processing cycle and a highly developed safety technology.

THE MOST EXPENSIVE SUBSTANCE ON EARTH

Just in case you were wondering, diamonds are far from the most expensive substance on Earth. Antimatter, with a price tag of 62.5 trillion dollars per gram has that distinction, given the costs of making it, and the difficulty of storing it. In contrast a one-gram high-quality diamond would cost roughly $300,000, or 200 million times less. Incidentally, if one single gram of antimatter were to completely annihilate one gram of matter, it would create an explosion roughly equivalent to three Hiroshima-size nuclear bombs.

How to Get a "Kinky" Spectrum

As was explained in Chapter 1, each of the three neutrino flavors, including the electron neutrino, is known to be comprised of three different neutrino masses. Nevertheless, experiments to find the electron neutrino mass almost always have fit the observed beta spectrum to a single mass. A single mass is used because in the standard neutrino model the three masses making up the electron neutrino have a tiny separation – see Figure 1.11 in Chapter 1. In fact, as that figure shows, unless an experiment could measure m^2 to better than 0.0025 eV^2, there would be no hope of observing the individual masses. This use of a single mass here is reminiscent of conventional analyses of SN 1987A neutrinos, which made the same assumption. You may recall from

Chapter 3 that by dropping the assumption of a single effective mass, and letting the SN 1987A data *speak for themselves,* I was led to the 3 + 3 model with its three very unequal masses. Given the three very unequal masses of this model, we obviously *cannot* describe the shape of the beta spectrum using a single effective mass. Instead, the appropriate procedure is to form a weighted sum of three spectra – one for each of the three masses with some appropriately chosen weighting factors.

To understand better how the spectrum shape is affected by the presence of several distinguishable masses, consider the following *fictitious* example in which half of electron neutrinos had a mass of 1 eV, and half had a mass of 2 eV. In that case, we would need to combine, that is average, the two spectra in Figure 5.7 for those masses, one of which ends at –1 eV, and the other at –2 eV. The resulting spectrum is found by simply taking the average height of the 1 eV and 2 eV spectra at each energy. This resultant spectrum, which ends at –1 eV has a *kink* at –2 eV – see dashed curve. Note that in this fictitious example with 50% of the neutrinos having a 1 eV mass and 50% a 2 eV mass results in more noticeable "kinkiness" than if the two contributions had been unequal. For example, the kink observed in Figure 5.7 at –2 eV would have been much less noticeable if the two contributions to the spectrum had been weighted 99% from one mass and only 1% from the other. In that case, the combined spectrum would look virtually indistinguishable from that for the more heavily weighted mass, and the spectrum would have no visible kink.

The preceding example explained how the presence of two masses (1 eV and 2 eV) would affect the spectrum by creating a noticeable kink, assuming the two contributions from each mass were not too unequal. The situation for three masses is a simple extension of this idea. Thus, if you did an experiment and you found the observed spectrum went to zero at an energy $-E_1$ (defined relative to zero at 18.574 keV), and that it also showed kinks located at energies $-E_2$ and $-E_3$. Such a result would imply the presence of three neutrino masses: $m_1 = E_1$, $m_2 = E_2$, and $m_3 = E_3$.

The Beta Spectrum Shape for a Tachyon

In the preceding examples we considered the impact on the spectrum of two or three *real* values of the neutrino mass, so now let us consider how an imaginary mass would affect the spectrum. We shall assume that the same so-called phase space function applies independent of the sign of m^2. In this case, the spectrum height gradually drops to zero for a tachyon, reaching zero at the same endpoint as a zero mass neutrino. We also assume that for a tachyon the spectrum height remains zero for energies above E_0, which some theorists deny. Even though the endpoint of the spectrum is assumed to be the same for a tachyonic neutrino as for one having zero mass, the shape of the spectrum near the endpoint differs in the two cases. Applying the same phase space function for the two cases, a zero-mass neutrino yields a parabola, as noted earlier, while an imaginary mass neutrino yields a linear decline in the spectrum close to the endpoint. You can see the parabolic decline case in Figure 5.7 labeled m = 0. In the special case of three neutrinos where two have real masses and one has an imaginary mass, as in the 3 + 3 model, we would expect the spectrum to show two kinks (at the locations of the two real masses), followed by a linear decline to the endpoint, for the imaginary mass. However, as we shall see later, these above results for the spectrum shape need to be modified if we are considering the KATRIN experiment because of the way it records data.

Neutrino Mass Experiments

Before we discuss the current KATRIN experiment, we first consider the earlier work involving about a dozen neutrino mass experiments, beginning in the late 1980s. The most recent and most precise pair of these experiments were by the Mainz Collaboration based in Germany and the Troitsk Collaboration based in Russia. Based on the shape of their observed tritium beta decay spectra these experiments reported consistent upper limits on the electron neutrino mass of 2 eV. Given the great importance to particle physics in learning the true value of the neutrino mass, and not simply an upper limit, finding a way to measure it with higher precision has been an important priority. The 20-year gap between the current KATRIN experiment and previous ones reflects the difficulty of reducing all the sources of systematic error, so that if an unexpected result (that is, an "anomaly") arises one can be confident that it is a real physical effect.

This last comment is especially relevant for neutrino mass experiments, in which an interesting anomaly has consistently appeared. All but one of ten experiments through the late 1990s have reported negative m^2 values, that is, imaginary masses, which in some cases, were well outside the experimental uncertainty. The experiments since then have had much smaller uncertainties, but they continued to report negative best values of m^2.

An Embarrassing Episode

Those earlier negative m^2 values almost certainly were the result of systematic effects, and they did *not* indicate neutrinos were tachyons, *contrary to what I once wrongly believed.* What did those earlier experiments reveal about the possible presence of three different mass neutrinos, as in the 3 + 3 model? In a 2016 paper I claimed that the Troitsk and Mainz data supported the 3 + 3 model with its three specific widely separated masses much better than they fit a single, small effective mass of 2 eV or less. That paper turns out to have been based on a regrettable error! I discuss it here because it shows how easily wishful thinking can affect our interpretation of data, *especially* when the data fit our preferred model very well.

The Troitsk Collaboration, which published their results in 1999, had been plagued with an unexplained kink near the end of the spectrum referred to as the "Troitsk anomaly." Their anomaly was an embarrassment to the Troitsk Collaboration, which they finally eliminated in a reassessment of their data in 2012 – fully 12 years after their data were first published. This 2012 paper by Troitsk did not explain the anomaly, but it instead eliminated it by excluding about a third of their data as suspect, although they provided no graph of the spectrum after this exclusion. The practice of data deletion, done to achieve a good fit to a favored hypothesis, is normally considered a dubious procedure, even if the deletion can be justified on a post hoc basis.

Recall that in Chapter 3 we showed how data from the 1987 supernova led to an unconventional 3 + 3 model that included an imaginary mass for one of three active-sterile neutrino pairs, and the other two pairs had masses of 4 and 21 eV. Those values are much heavier and further apart than what is normally assumed in the standard neutrino mass model in which there are three nearly identical masses. What would the 3 + 3 model predict for the shape of the spectrum near the endpoint? Based on the earlier discussion, given two $m^2 > 0$ masses of 4 eV and 21 eV, the model would predict a pair of kinks occurring at 21 eV and 4 eV before the spectrum endpoint. Neither of these two predicted kinks would appear consistent with the one that Troitsk found at around 10 eV before the endpoint. On the other hand, with an appropriate shift in the location of the kink relative to E_0, one finds a very good fit of the Troitsk data to the 3 + 3 model.

Of course, making an arbitrary adjustment to the energy scale to achieve a good fit might seem highly questionable. On the other hand, in the 2012 Troitsk "reassessment paper" referred to earlier in which they eliminated their unwanted anomaly they also noted that its location had in fact been about 20 eV from the endpoint, and not 10 eV as they had originally claimed. On learning of this shift, I became very confident that the Troitsk anomaly in their original data set could be nicely explained by the 3 + 3 model, even if they had later eliminated their anomaly by excluding some of their data.

In summary, I had been eager to find evidence for my 3 + 3 model, and I managed to convince myself that the data from the Troitsk experiment strongly supported it based on a kink in their spectra occurring about 20 eV from the endpoint. I even found further support for the existence of such a kink in two other high-precision experiments done at Mainz and Lawrence Livermore Lab. Sadly, I was mistaken. The observed kink found near –20 eV in the three sets of data had another explanation discussed later. Thus, those pre-KATRIN data in fact did not support the 3 + 3 model to the exclusion of a single small neutrino mass. As we shall see, however, those earlier experiments did not contradict the model either; they just were not sensitive enough to properly test it.

The KATRIN Experiment

KATRIN is the felicitous acronym for the **KA**rlsruhe **TRI**tium **N**eutrino experiment being performed at the Karlsruhe Institute of Technology (KIT). This lab is almost unique in the world, since only Japan operates a research laboratory with a similar tritium inventory, where the material is delivered to experiments, recovered, and reprocessed for further use. It is interesting that the acronym for the experiment, KATRIN, also happens to be the German variation of Katherine, whose Greek cognate means "pure." This name makes it a fitting choice for an experiment hoping to measure the neutrino mass free from the corrupting systematic effects that have plagued many earlier attempts. The KATRIN Collaboration is an international group of 150 scientists, engineers, technicians, and students from 12 institutions in Germany, the United Kingdom, the Russian Federation, the Czech Republic, and the United States. Several people in the group have worked on the earlier Mainz and Troitsk experiments, some when they were students. It has taken almost 16 years from the first design to the completion of the apparatus because of the many new technologies that had to be developed to achieve the designed precision. KATRIN started taking data in June 2018, following a long period of construction, preparation, and testing, with the goal of minimizing systematic uncertainties. Such attention to detail was considered essential if the experiment is to achieve its design goal of either measuring an actual value for the neutrino mass, if it was larger than 0.35 eV, or otherwise improving the upper limit to 0.2 eV. Achieving this goal should require 5 to 6 years, which would involve roughly 3-years of actual data-taking divided into a series of runs or "campaigns." Achieving an upper limit of 0.2 eV would represent a factor of ten improvement over the earlier 2 eV limit set by the Mainz and Troitsk experiments.

KATRIN is certainly not a table-top experiment. Its main component is a 200-ton spectrometer shown in Figure 5.8 being *very* carefully hauled through the town of Eggenstein-Leopoldshafen on the final leg of its journey to the site of the experiment. The spectrometer was built in the town of Deggendorf, 350 km from Karlsruhe. However, due to the spectrometer's size, land transport the whole way was impossible. Instead, the device was shipped down the Danube to the Black Sea, through the Mediterranean and Atlantic Ocean, through the English Channel, and then up the Rhine to Karlsruhe – a 8600 km detour (see Figure 5.9), which limited the land travel portion to the final 7 km from the Eggenstein-Leopoldshafen docks to the laboratory.

FIGURE 5.8
The KATRIN main spectrometer being moved through the streets of Eggenstein-Leopoldshafen.

Source: Courtesy of Gabi Zachmann, KIT

The main spectrometer and other components of the experiment are shown in Figure 5.10. Gaseous tritium emanates from the source at the left end of the apparatus at less than one hundredth atmospheric pressure. The tritium gas and the beta decay electrons they produce travel through a long pipe toward the main spectrometer, which is where the electrons have their energy measured. One unique aspect of KATRIN is its very low background, which makes it much easier to determine the precise spectrum endpoint

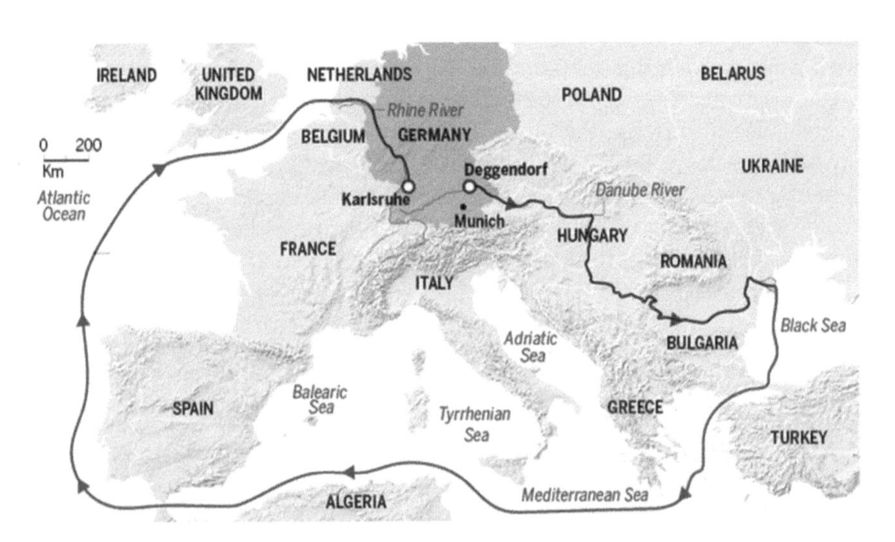

FIGURE 5.9
The route taken by the spectrometer from Deggendorf, Germany to KIT.

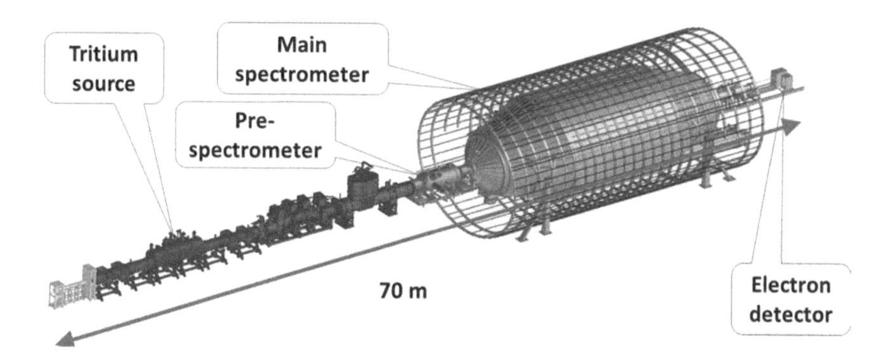

FIGURE 5.10
The main components of the KATRIN experiment.

whose value was not well known in prior experiments. The low background requires that the spectrometer be kept as free of tritium gas as possible. In fact, the pressure inside the spectrometer is a mere 10^{-14} atmospheres which is comparable to that on the lunar surface. Since the gas leaves the tritium source at less than 0.01 atmospheres, as it moves down the pipe towards the main spectrometer, its pressure drops by a factor of a trillion.

Measuring the Electron Energy in the Spectrometer

A cross section of the main spectrometer is shown in Figure 5.11. The voltage of the spectrometer electrodes causes beta decay electrons to ***de***celerate from the entrance point (labeled source in Figure 5.11) and then accelerate after the midpoint to the detector on the right, which they only reach if their energy exceeds some preset value.

Measuring the energy of electrons emitted in beta decay is here much like finding the energy of balls by rolling them towards a hill of adjustable height to see if they make it over the top. By making the height of the hill adjustable, one can measure the number of electrons having above a series of "set point" energies.

It is important, of course, that the electrons during their passage through the spectrometer be kept away from the walls where they could be absorbed. This is accomplished by a magnetic field which causes electrons to spiral around the field lines depicted in Figure 5.11 as they travel through the spectrometer. Consider the three electron tracks shown in the figure. Track (a) is for an electron from the source that has enough energy to make it over the "top of the hill" and reach the detector. Track (c) is for an electron that does not have enough energy (i.e., "rolls back down the hill") and does not reach the detector. Track (b) is for an electron that originates from a tritium beta

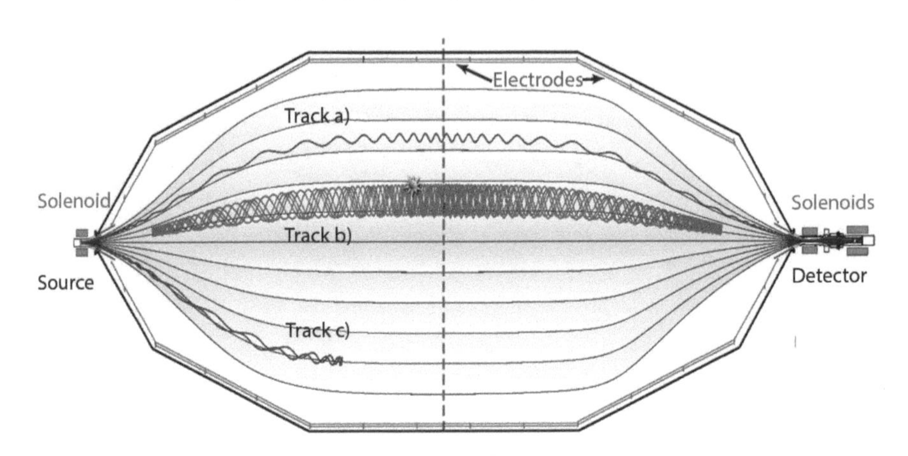

FIGURE 5.11
Cross sectional view of main spectrometer showing magnetic field lines and three particle tracks, a, b, and c. Drawing by John P. Barrett et al. CC A 3.0 Unported license

decay inside the spectrometer. If this electron were to reach the detector it would be a background event.

The energy resolution of the KATRIN spectrometer is better than 1 eV, so that the experiment can recognize very fine features in the spectrum with this precision. The three advantages of KATRIN discussed here (stable intense tritium source, low background rate, and good energy resolution) all result in very small statistical and systematic uncertainties, which are the keys to finding a reliable shape of the spectrum near the endpoint, and hence a value for the neutrino mass. There are, however, some unusual features in the way the experiment records the observed spectrum. First, KATRIN records what is known as an *integral* spectrum, that is, it counts the number of electrons having *above* the series of set point energies. In contrast, all the spectra we have so far depicted in graphs show the number of electrons in small intervals *at* each energy. A second unusual feature is that the set point energies at which the spectrum is recorded are not equally spaced. Instead, their values and the number of hours to record data for each energy were chosen in advance to minimize the uncertainty in the fitted neutrino mass. A third feature in the way data is recorded is that they were taken in blinded mode.

Taking Data with Your Eyes Closed

Acquiring data while blinded eliminates, or at least reduces, observer bias, because none of the experimenters knew what the results were as they were coming in, and the data were only unblinded after their analysis was completed. Without a blinded analysis one can fall into the trap of making data cuts or selections that make them fit better to some favored hypothesis by eliminating data that does not fit. Recall the example of the Troitsk anomaly, which was eliminated after excluding about a third of the data many years after the data were recorded.

A very early use of blinded analysis in physics goes back to 1933 in an experiment that measured the charge to mass ratio of the electron. This quantity, usually written as e/m, was proportional to the angle between the electron source and detector. Blinding was done by having someone choose a random angle for this quantity, which was hidden from the experimenter, who only measured its value when the result was ready to be published. In medicine one normally finds *double* blinding, where neither patient nor doctor knows whether a placebo or real drug or medical procedure is being administered. Double blinding is now a routine practice in medicine, having come into first use in 1948. Amazingly, the practice of blinding to eliminate bias in medical studies was suggested as far back as 1662 by John Baptist van Helmont, a Flemish chemist, physiologist, and physician, but it was not utilized in medicine for nearly four centuries. In physics only single blinding is needed because we do not expect electrons in beta decay to be influenced by any belief *they* may have about the shape of their spectrum, unlike the placebo effect that can affect patients' outcomes in drug trials.

Dealing with Controversial Results

The KATRIN experiment started taking data in June 2018, but not until March 2019 did it began taking data at the full planned tritium flow rate; the earlier data were for calibration purposes. The plan was to take data until July 2019 in blinded mode, and only then unblind the results. As a very impatient person, the wait during this period was excruciating for me, since I then believed that once the initial results were unblinded it would become completely clear that my 3 + 3 model with its tachyon neutrino would either be confirmed or refuted, with virtually no room for ambiguity. I therefore reached out by email to Guido Drexlin, co-spokesperson of KATRIN, about two weeks after their unblinding to ask about their results, but I received no answer. Drexlin had in the past responded to messages, so I was unsure if the lack of response this time reflected his heavy workload or something else. It occurred to me that if my model seemed to be validated, they probably would like to keep this fact secret until they could be certain this was not some kind of mirage. They even would want to go ahead and write the paper on the results and have every statement in it reviewed in excruciating detail by all team members before making it public – a process that might take considerable time. The physics world still vividly remembers what happened when the OPERA Collaboration discovered their infamous anomaly in 2011, which seemed to show neutrinos were faster than light (FTL) tachyons. In that case, the group was essentially forced to announce their result before everything could be checked, because word had begun to leak out. The anomaly eventually was traced to experimental flaws, and after correction, the neutrino speed was consistent with that of light within uncertainties – their anomaly evaporated.

It is instructive to recall a bit more about the history of this episode, and how a potentially revolutionary discovery was handled. When the OPERA group first found its FTL neutrino anomaly there was considerable disagreement within the group on how to proceed, with some members of the 170-person team arguing that in view of its controversial nature the result should be kept secret longer, while others disagreed, stressing the value of having others outside the group study it, while the team continued to double check everything. A few in the first category declined to have their names on the original paper reporting the anomaly. These tensions within the group eventually led to a vote of no confidence in Antonio Ereditato the group spokesman. Although the motion failed to pass, Ereditato resigned the day following the vote, as did the physics coordinator Dario Autiero. They resigned to calm the tensions within the group, which apparently existed before the anomaly was found. Ereditato has insisted that what they did in announcing the result conforms to the way science should be done, and that it should not be regarded as a mistake. In his words, "Even when they are particularly unexpected or uncomfortable, findings must be made public, entailing scrutiny by the scientific community. Never did I or any of my colleagues at OPERA talk of a discovery or a final result."[2]

Ereditato is an extremely well-regarded neutrino physicist, having over 1100 scientific publications. It is indeed a shame that he will probably be remembered primarily for his role in the OPERA neutrino anomaly, which has been unfairly characterized by some media reports as a blunder. It may have been a blunder in the sense of there being a mistake in the experiment arising from a loose cable and a clock that ticked too fast, but they never claimed their initial anomaly was the result of superluminal neutrinos. The whole episode does, however, illustrate the danger of announcing a result to the media before the paper has been accepted by a refereed journal.

My Overactive Imagination

If the KATRIN people were sitting on a major discovery, especially one that involved neutrinos being tachyons, I was sure that they would do everything in their power to keep their results secret for as long as it took to check everything in the experiment, and not follow OPERA's path. Anyway, when I contacted Group leader Drexlin, I expressed my understanding that they would probably want to keep the results secret for a while if they validated my model. However, I also asked that they please let me know if their results were *not* consistent with my model. Phrasing my request to Drexlin in those negative terms was not a matter of sophistry on my part, rather I had then foolishly believed that my model could be ruled in or out just from a visual inspection of the observed spectrum.

As discussed earlier, I had believed the prominent kink at –21 eV that Troitsk had observed was due to one of the 3 + 3 model masses. It therefore seemed to me that if KATRIN did *not* see this same kink in their data it would not compromise their secrecy to simply inform me – "Sorry, Bob – it looks like your model is dead." In the past my inquiries about the experiment were usually answered promptly, so when after a few weeks of receiving no response from Drexlin, I began to imagine that in this case no news might be very good news. With each passing day of no email from Drexlin, I became more confident that I might soon get some very good news. In other words, not for the first time, I got carried away with my wishful thinking.

During this waiting period I was also aware that, while the KATRIN group treated me very courteously during my earlier visits to their lab, they probably did not believe my model with its tachyon neutrino could possibly be correct. In fact, in previous papers they had referred to the $m^2 < 0$ effective masses seen in earlier experiments as being "unphysical," and they had even written papers that successfully explained this anomaly in other ways. Their skepticism about the negative mass square values reported in earlier experiments being due to systematic effects and not real tachyons was almost certainly well-justified. However, I believed the situation was now quite different. If the KATRIN data gave a good fit to the 3 + 3 model (which included a tachyon) it would be far more difficult to explain away such a result as being the result of systematic effects. Not only had the experiment gone to great pains to eliminate all known systematic effects, but the 3 + 3 model made *three* specific predictions not just one about the beta spectrum near its endpoint due to the presence of three separate masses.

By the way, when experimenters try to explain away strange features appearing in their data that conflict with one's current understanding of physical law they are engaging in a legitimate activity. On the other hand, in doing so, they must not to "throw the baby out with the bath water," or else they will never discover novel changes in physical law. I believed that in this case, despite many physicists' disbelief in tachyons, if the three features predicted by the 3 + 3 model were indeed present in their data, the evidence would be sufficiently robust that it would withstand the best efforts to make it go away.

A Big Letdown

In September 2019, some weeks after unblinding their data the KATRIN group finally submitted a paper for publication. To my disappointment these results, which were based on their initial period of 33 days of data-taking, did not seem to support my 3 + 3 model prediction. Instead KATRIN found a spectrum that was consistent with neutrinos having a *single* mass of 1.1 eV or less. This first release of the KATRIN results marked the start of a period of two or three weeks during which I tried to find some way that might make their results consistent with the 3 + 3 model. I have always been suspicious of theorists who would not take "no" for an answer, and who tried to concoct some way their model was correct after an experiment conflicted with its predictions. However, I have also regarded it as an even worse mistake to give up on one's model prematurely, before making every possible attempt to find some plausible explanation as to why the model's predictions seemed to disagree with the results of an experiment. My initial attempts were not promising in this regard, but after many false starts I finally discovered the very simple answer.

Consistency of KATRIN Initial Data with 3 + 3 Model

The start of an explanation was the realization that the kink seen near –20 eV in the beta spectrum in the earlier experiments was not an essential feature of the 3 + 3 model, but rather a contingent one that depended on what fraction of the spectrum was contributed by the 21 eV mass. Thus, if the spectral contribution from the 21 eV mass were made very small, no observable kink would be seen at –21 eV. This idea was earlier explained in a fictitious example of having 99% of the spectrum from a 2 eV mass and 1% from a 1 eV mass. Of course, for this explanation to be correct the noticeable kink observed in those earlier experiments' spectra near –20 eV would need to have had some origin having nothing to do with the 3 + 3 model. In fact, there was an effect that could yield such a kink there, namely something called "final state distributions." This effect refers to the alteration of the beta spectrum near the endpoint due to excitations of a tritium molecule after the nucleus has undergone beta decay. By failing to take this effect into account in my analysis of pre-KATRIN experiments, I had fooled myself into believing those results presented evidence for the 3 + 3 model. The spectrum the KATRIN group used to fit its data *did* include the final state distributions, which accounted for most of the "kinkiness" in the spectrum seen near –20 eV. However, recall that KATRIN measures the *integral* spectrum. When displaying the integral spectrum, the difference between the 3 + 3 model spectrum and that for the

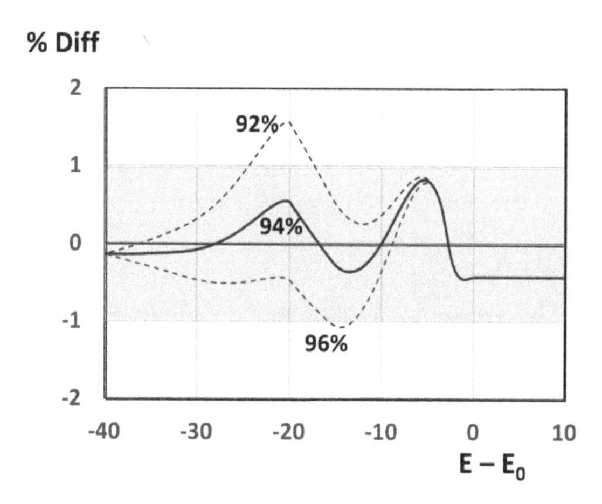

FIGURE 5.12
Percent difference between the integrated spectra for the 3 + 3 model and the standard single mass one for three different contributions from the 4 eV mass. 94% curve is the best fit.

standard single mass shows up as bumps (*not* kinks) located near energies of –4 eV and –21 eV – see Figure 5.12.

Despite having lost the support for the 3 + 3 model based on the pre-KATRIN experiments I continued to believe in its correctness, because the pre-KATRIN experiments did not contradict it – they just were not sensitive enough to test it. But what exactly would be required for the first results from KATRIN to be consistent with the model? If we calculate the expected shape of the integrated spectrum for the 3 + 3 model and the standard one with a single small mass, it turns out that consistency between them can be achieved, but *only* for a small range of possible values of the contributions to the spectrum from each of the three masses in the 3 + 3 model. In particular, the two models are consistent *only* if 94 ±2% of the total spectrum was from the 4 eV mass, 6 ±2% from the 21 eV mass, only a tiny contribution (0.094%) from the tachyonic ($m^2 < 0$) neutrino. Figure 5.12 illustrates why only a narrow range of contributions from the 4 eV mass makes the 3 + 3 model consistent with the standard one. As can be seen, when the contribution from the 4 eV mass is 94%, the curve lies entirely within the ±1.0% width of the grey band. However, when that contribution either rises to 96% or drops to 92% we find the (dashed) curves lie outside the grey band at some energies, and hence the difference between the two models becomes much greater – in fact, it is too large for both models to be consistent with the data.

In summary, as I noted in a 2019 paper, given the limited number of events in the first release of KATRIN data, their observed spectrum could be made consistent with *both* the standard small neutrino mass and the 3 + 3 model, but only for a very narrow range of possible contributions from the three masses in the model. However, as KATRIN accumulates more data, the size of the statistical uncertainties will shrink, and as a result, the data should be consistent either with the spectrum for the 3 + 3 model or that for a single small effective mass, *but not both.*

A Stay of Execution

I concluded that I had essentially gotten a reprieve, or a "stay of execution" for my model, which effectively was still on death row. During this time, I reached out to Diana Parno, one of the leaders of the KATRIN group overseeing the data analysis, who assured me that they were taking my 3 + 3 model seriously, and that having read my 2019 paper, she was aware of my assertion that it indeed fit their first results at least as well as a single mass did. Having assured myself that there was a way for my model to be viable, and that its fate could be sealed once new data were unblinded, I could only wait and occupy my time with other things, including writing this book and reading. My wife Elaine and I are members of a book club, and not entirely by coincidence one of our monthly selections was the 1980 sci-fi novel *Timescape,* which I referred to earlier. This tale of scientists using tachyons to send a message back in time to avert an environmental disaster is one of my favorite sci-fi novels. It felt downright weird to be reading about tachyons in a novel, given my hope that in a matter of months I might learn they were not just science fiction. During this waiting period I vowed to try to restrain my overactive imagination, and simply wait patiently for KATRIN to complete their analysis, which this time I believed should be a definitive test of the 3 + 3 model in perhaps a year.

Hearing the Grim Reaper's Footsteps

My original plan in writing this book was to finish it all except for KATRIN's results, and then if they did fulfill my model's predictions only then seek out a publisher. In fact, I expected that if the model were confirmed many publishers would seek *me* out, given the revolutionary implications of some neutrinos being tachyons. As already noted, I was confident that those KATRIN results would either confirm or doom the model within a matter of perhaps a year. Thus, as an 83-year-old in reasonable health, I had been confident that I would live to learn the outcome, and see this book published. The Coronavirus pandemic, however, made me reassess that optimistic view, especially since I am in a high-risk group based on my age and gender. I, therefore, thought it would be prudent to make arrangements to complete this book, in case I am not around to do so. Alan Chodos, who has done extensive research on neutrinos as tachyons, has very kindly agreed to deal with any final editing of the book in the event of my death or incapacitation. Alan was also responsible for encouraging me to write this book in the first place, regardless of how the KATRIN experiment turns out.

KATRIN and Tachyons – Six Possibilities

I was recently asked about the fate of the tachyonic neutrino hypothesis if the KATRIN results were to disprove the 3 + 3 model. Would I accept that tachyonic neutrinos do not exist based on such a result, or would I stubbornly hang on to a discredited hypothesis? My answer is that while the 3 + 3 model would indeed be dead in such a case, the possibility of some neutrinos being tachyons would be alive and well, regardless of KATRIN's results. In fact, with respect to the notion of some neutrinos being tachyons, one can imagine six possible outcomes of the experiment as listed below. The first four outcomes assume that the 3 + 3 model with its tachyon neutrino is invalidated, and the standard neutrino mass model with three neutrinos having nearly the same mass is upheld.

1. **Smaller upper limit on m.** The least surprising outcome for most physicists would be if the experiment finds a smaller upper limit on the electron neutrino effective mass. This result would still mean the mass could be imaginary, or m^2 could be negative, within some experimental uncertainty. This is what was found based on KATRIN's initial published results.
2. **Actual mass determination having positive m^2.** A second outcome would be an actual value for the effective mass of the electron neutrino, having positive m^2, which should be possible after 5 to 6 years provided the mass exceeds 0.35 eV. Such an outcome, however, would be extremely surprising given that it directly conflicts with the result from cosmology that the sum of the three neutrino masses is less than 0.1 eV. This outcome would rule out the possibility of the *electron* neutrino being a tachyon. However, it would provide no information about whether the other two neutrino flavors (muon and tau) might be tachyons.
3. **A negative m^2 short of the 5-sigma discovery level.** A third outcome is a negative m^2 mass for the electron neutrino at perhaps the level of three or four standard deviations from zero. Such a result undoubtedly would encourage KATRIN to extend their planned data-taking period, and other direct mass experiments to accelerate their efforts. It might also drive KATRIN to seek systematic effects that might have been overlooked.

4. **Discovery of a negative m^2 electron neutrino.** A fourth outcome is a negative m^2 mass for the electron neutrino at the 5-sigma discovery level. Apparently, the sensitivity of KATRIN is such that this might be achievable if the value of m^2 were more negative than -0.11 eV^2, which by a striking coincidence also happens to be the exact value I have predicted for the electron neutrino effective mass in a 2015 paper. This prediction was based on six different kinds of observations, and it was independent of the 3 + 3 model.
5. **Validation of the 3 + 3 model.** The most revolutionary outcome would be the validation of the 3 + 3 model which includes a tachyonic neutrino having an approximate $m^2 \sim -0.2$ keV^2. This result is consistent with KATRIN's initial published results, as noted previously. However, given the revolutionary implications, one might imagine that the Collaboration might want to delay announcement of such a result for as long as possible, while accumulating more data.
6. **Ambiguity between 3 + 3 model and conventional one.** Although KATRIN's first results have exactly this ambiguity, one or the other hypothesis should become favored as more data is collected. Given the difference in spectral shape between the 3 + 3 model and a single effective mass illustrated in Figure 5.12, it is not possible for an ambiguity to remain indefinitely as the data error bars shrink with increasing numbers of events.

Fitting an Elephant or a Whole Herd

When KATRIN published their initial results in 2019, they showed a fit to the observed tritium beta spectrum using four free parameters. The meaning of those parameters is illustrated in Figure 5.13: (1) the signal amplitude at some specific energy, "Sig," (2) the spectrum endpoint E_0, (3) the background count rate, "Bkgd," and (4) the neutrino mass square, m^2. The mathematician John von Neumann once famously said: "With four parameters I can fit an elephant, and with five I can make him wiggle his trunk."[3] Von Neumann's humorous remark was intended as a caution against using more free parameters than justified, that is, engaging in *overfitting*, when fitting data to some model. The main danger of overfitting is that you could obtain an excellent fit to virtually *any* desired shape, so finding a good agreement between experiment and theory is not convincing. For example, taking a ridiculous extreme case, if we had a plot with 12 data points, fitting the data with 12 free parameters would result in a curve passing through all the points. It then would be absurd to interpret the four peaks in the plot as having any significance whatsoever, especially given the size of the error bars in Figure 5.14. As will

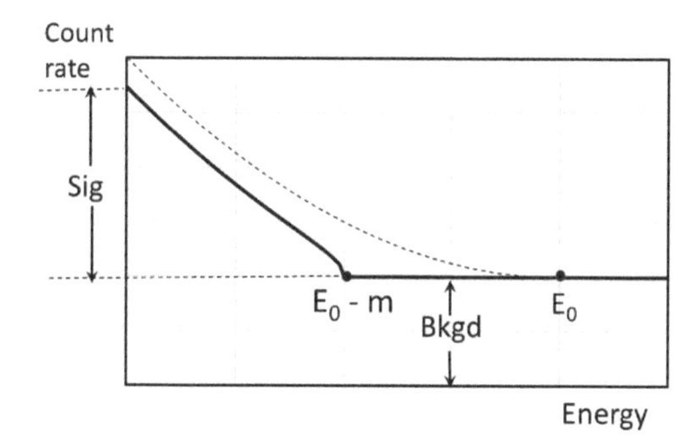

FIGURE 5.13
The four parameters used to describe the spectrum shape for a single effective neutrino mass, m. The dotted curve is for the case m = 0.

be seen, there is a second more subtle danger to overfitting: not only could you fit an elephant with too many free parameters but you could also make unwanted elephants disappear (see Figure 5.15). Contrary to von Neumann's observation, KATRIN in its 2019 paper, was ***not*** engaging in overfitting, since in this case, four adjustable parameters would be exactly the right number to use, assuming that the endpoint energy E_0 is not precisely known, and only one neutrino mass is present.

As discussed earlier, those initial 2019 results gave a good fit to a single small effective neutrino mass, but they also fit my 3 + 3 neutrino model with its three different masses – one being a tachyon. In fact, they slightly favored the 3 + 3 model. I had, therefore, been eagerly awaiting their second data release which might either kill or confirm my model. These new data taken in a second run were presented to the world at the March 2021 meeting of

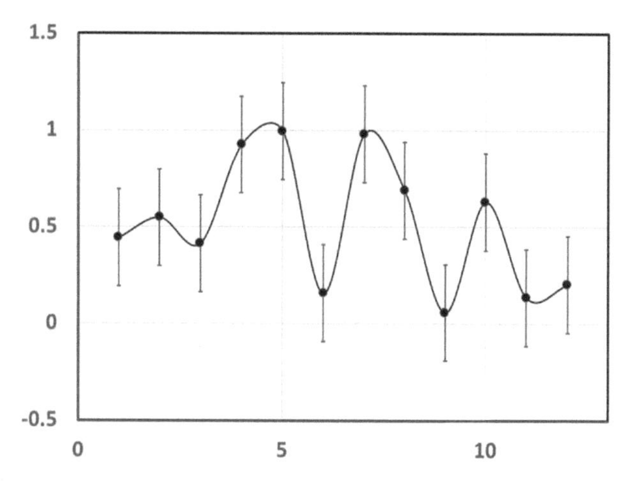

FIGURE 5.14
An extreme example of overfitting given the size of the error bars.

FIGURE 5.15
"With our 31 free parameters, they'll never spot us."

Source: Photo by hansen.mathew.d from Shutterstock

the German Physical Society. At that meeting a KATRIN spokesperson displayed a graph that showed an excellent fit to the spectrum combining the data from the two runs. My heart sank when I first viewed this graph, because there was simply no sign of any small bumps occurring near the energies –4 and –21 eV, which my 3 + 3 model required – see Figure 5.12. However, on closer inspection I noticed that KATRIN had used not 4 free parameters to fit their spectrum, but rather 31 of them, which would be enough to fit (or perhaps hide) a whole herd of elephants! For example, let us suppose the data showed a strong preference for the 3 + 3 model, meaning it showed two small bumps, that is, excess numbers of events (on the order of 1%) near the energies –4 eV and –21 eV in Figure 5.12. As we shall see, overfitting could make those bumps be statistically insignificant. The absence of any fit to the spectrum using four free parameters was particularly surprising, because that is what was done in the Collaboration's first paper in 2019. Moreover, two months after the 2021 meeting, KATRIN posted their results in a paper on the e-print archive, and now the number of free parameters used in their fit had increased yet again to 37. The size of the elephant herd had continued to grow.

To understand how KATRIN came up with 37 free parameters, note that there are 148 pixels on the detector surface which can be divided into 12 circular rings. Given the cylindrical geometry of the spectrometer, it is plausible that the data for each ring on the detector might have a different signal strength, endpoint energy, and background. These would yield $3 \times 12 = 36$ free parameters, plus one more for the neutrino mass. Roughly speaking, you can think of the data for each detector ring being fit to obtain a separate measurement of the neutrino mass, and the final mass is obtained from a weighted average of those 12 values. Thus, there is some rationale for using as many free parameters as 37, and by doing so, KATRIN was almost certainly not deliberately hiding an interesting result.

Hiding Elephants

Not only can using too many free parameters fit imaginary elephants, but it can also hide real ones. The first possibility (a false positive) is easy to understand. Thus, it is clear how using too many free parameters can yield spurious features in the spectrum. Just remember the example of the ridiculous four peaks in Figure 5.14. However, it is much less obvious how using too many parameters could hide real features (produce a false negative). Such masking or hiding real features can occur because systematic departures from the single mass spectrum might be too small to be statistically significant for the fit to the data for any one ring of pixels.

To be specific, suppose the data for every one of the 12 rings showed one to two standard deviation excess counts at the predicted locations for the two bumps in the 3 + 3 model. Such deviations would be small enough to allow a good fit to a single mass spectrum for each ring, and hence the anomalies could go unnoticed. However, if the data had not been divided into 12 separate rings, and a single four parameter fit were done for the data from all rings combined, the excess where the two bumps were predicted could easily have been more than 5 standard deviations.

As it turns out, however, KATRIN did do a four-parameter fit to their new data, as they did in their earlier 2019 publication, but it was just not highlighted in their 2021 paper. Moreover, this fit to the new data apparently gave consistent results with their 37-parameter fit. On the other hand, I was very encouraged to observe that just like their fit to the first dataset the fit to the new data showed a two-sigma excess number of counts very close to –21 eV, which is one of the two locations where the 3 + 3 model predicted small bumps. These excesses gave me great hope that my model with its tachyonic neutrino remained alive and well, at least until the KATRIN people accumulate much more data and they get around to doing a proper 3 + 3 model fit to all of it, hopefully not divided up into individual rings.

Summary

Weighing the neutrino has been a long-time goal of particle physics that has proven to be exceptionally difficult, given the smallness of their masses. Of the three neutrino flavors, the muon neutrino mass can be found in principle by looking at individual events in which a pion decays, but for the electron neutrino a statistical distribution of events is needed. The shape of the beta decay spectrum near the endpoint can reveal the electron neutrino mass. Earlier beta decay experiments done by various groups only were able to set upper limits on the mass, and the most accurate ones before KATRIN by the Mainz and Troitsk groups lowered the upper limit to 2 eV. Experiments normally assume the three neutrino masses have nearly equal values, based on the standard neutrino model, so that only a single effective mass for the electron neutrino can be used. On the other hand, my 3 + 3 model of the neutrino masses, including one mass which is imaginary, fits those earlier data as well as the usual single effective mass, even though in a 2016 paper I had mistakenly claimed they fit the 3 + 3 model better.

The KATRIN experiment has much greater sensitivity than earlier ones. This experiment's first release of data in 2019 lowered the upper limit on the neutrino mass to 1.1 eV. However, it also gave a good fit to the 3 + 3 model as well, as I showed in a 2019 paper, but only for a very narrow range of contributions from each of the three masses in the model. I initially expected that after another year of data-taking, KATRIN would have enough data to resolve this ambiguity, and either confirm or disprove the 3 + 3 model. Unfortunately, however, the situation is still uncertain two years after the first data release, although there are encouraging hints in their fits to the data for the first two runs.

References

1. Brown, Dan. *Angels & Demons*, New York, NY, Pocket Books, 2000.
2. Ereditato, Antonio. "OPERA: Ereditato's Point of View." March 30, 2012. https://www.lescienze.it/news/2012/03/30/news/opera_ereditatos_point_of_view-938232/
3. Dyson, Freeman. A Meeting With Enrico Fermi, *Nature (London)* 427, 297, 2004.

6

Lessons Learned

Only one who attempts the absurd can achieve the impossible.

Message in a fortune cookie eaten by me on August 2, 2021[1]

A Bright Spot in the Darkness

When observations are used to try to uncover the nature's secrets we usually rely on a kind of backward reasoning. Let us see how the process works using the 3 + 3 model with its tachyon neutrino as an example. Table 6.1 identifies 15 pieces of evidence for the model published in papers between 2012 and 2019, only some of which have been discussed in previous chapters. A few of the key points from the table are noted below. First, an analysis of the neutrinos from supernova 1987A led to an unconventional 3 + 3 model of three active-sterile neutrino pairs, with one pair being tachyons, having an approximate mass squared $m^2 \sim -0.2$ keV^2. Later this tachyonic mass was found to match the value inferred from the Mont Blanc neutrino burst, but only on the weird condition that there existed an 8 MeV neutrino line in the spectrum for the supernova, something initially ruled out as "inconceivable." Subsequently, a model was discovered for creating such a line and the model was validated using galactic center gamma rays. Additionally, evidence for an 8 MeV line was found atop the background of 997 neutrino events recorded on the date of SN 1987A, thereby confirming the model. The basic logic here has the following form:

> Exotic model X predicts super-weird thing Y (an 8 MeV neutrino line). Evidence for Y is then found. X is thereby confirmed.

Any logician or statistician would laugh at this reasoning noting that just because X implies Y it certainly does not follow that Y implies X, but such backward reasoning happens routinely in empirical science. My favorite example is a dramatic development that occurred in early 19th century

DOI: 10.1201/9781003152965-6

TABLE 6.1

Evidence supporting the 3 + 3 model of the neutrino masses including a tachyon.

	Evidence	Comment	Ref
1	The SN 1987A neutrinos in 3 detectors fit one of two masses.	Assumes very small spread in neutrino emission times.	E(2012)
2	3 + 3 model: Above 2 masses yield same fractional Splittings → 3rd mass $m^2 = -0.2$ keV^2.	Assumes a $dm^2 \sim 1$ eV^2 seen in some oscillation experiments.	E(2013)
3	The 5 Mont Blanc neutrinos are consistent with having a common energy ~ 8 MeV.	No other SN model explains this observed constancy.	E(2018)
4	The computed tachyonic mass for the Mont Blanc neutrinos equal that in the 3 + 3 model.	A common mass for the Mont Blanc burst requires an 8 MeV neutrino line.	E(2018)
5	The 16.7 MeV Z′ Boson and 8 MeV cold dark matter X particles yield an 8 MeV neutrino line.	Z′ is a likely mediator between ordinary and dark matter.	E(2018)
6	The spectrum of gamma rays from the galactic center agrees with that for XX annihilation.	X particles would be expected to annihilate near the massive BH	E(2018)
7	The observed temperature and angular source size also agree with XX annihilation.	The calculated temperature is based on the low temperature solution.	E(2018)
8	The Kamiokande data on the day of SN 1987A yield an 8 MeV neutrino line.	The background was obtained by another dataset taken months later.	E(2018)
9	The 8 MeV neutrino line is found broadened by 25% agreeing with energy resolution.	The line was well hidden given the similar shapes of the background & line.	E(2018)
10	The 21 eV mass in the 3 + 3 model fits the observed dark matter profile in our Galaxy.	The tachyonic mass in 3 + 3 could be associated with dark energy.	C(2014)
11	The 4 eV mass in the 3 + 3 model fits the dark matter distribution in four galaxy clusters.	The smaller mass would be the dominant one for the larger structure.	C(2014)
12	Three beta decay experiments fit 3 + 3 model masses *as well as* they fit a mass m < 2 eV. Ref E(2016) wrongly claimed a "better" fit.	E(2016) failed to take into account effect of final state distributions – see discussion in E(2019).	E(2016)
13	Weights assigned to the 3 masses in the above fits correspond to a near-zero effective mass.	Despite the very small effective mass fits need to be a sum of 3 spectra.	E(2016)
14	Six observations suggest a value of $m^2 = -.011$ eV^2 for the neutrino effective mass squared.	A $m^2 < 0$ mass in the 3 + 3 model requires at least one $m^2 < 0$ flavor state.	E(2015)
15	The first KATRIN results fit the 3 + 3 model	Only for a very specific mass mixture.	E(2019) E(2021)

C (2014) is the paper by Chan & Ehrlich in 2014, E(2012) is the paper by Ehrlich in 2012, etc. See the web site for this book for more details on the references: Ehrlich.physics.gmu.edu

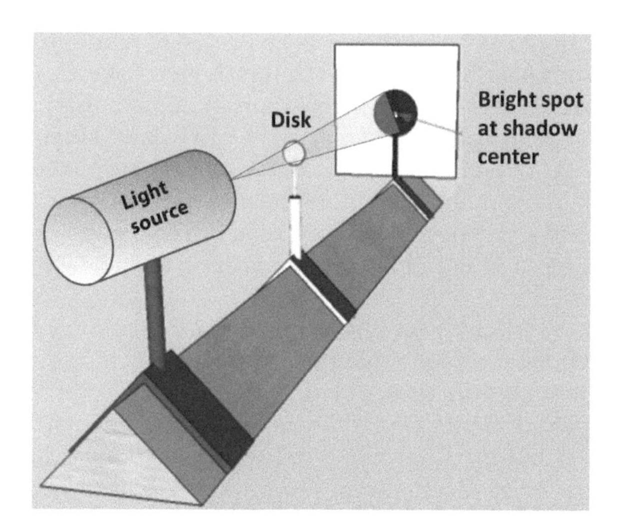

FIGURE 6.1
Arago/Poisson spot experiment.

France. At that time, most physicists accepted Newton's particle theory of light, over the alternative wave theory. The French scientist Siméon Denis Poisson, a skeptic of the wave theory, then showed that it predicted a bright spot right at the center of the dark shadow of a perfectly circular disk – see Figure 6.1.

Such a spot, of course, is not observed in everyday situations, and Poisson concluded that his completely absurd result disproved the wave theory. Francois Arago (who later became Prime Minister of France!) thought that just to be safe it might be worthwhile to check if Poisson's "ridiculous" prediction could be true. After performing a careful experiment, Arago found exactly what Poisson had calculated, which convinced most scientists that the wave nature of light was correct despite their earlier skepticism. After the bright spot was found, few physicists persisted in trying to accommodate this result within the particle theory. Moreover, no logicians are known to have objected regarding the clear fallacy involved here: just because the wave theory predicts a bright spot in the darkness, it does not follow logically that finding such a spot requires the wave theory to be correct.

Making the whole episode even more intriguing, it turns out that the Poisson or Arago spot had first been independently observed a century earlier. Two French scientists, Joseph-Nicola Delisle and Giacomo Filippo Maraldi separately reported the spot, a fact that was apparently unknown to Poisson when he made his calculation. Incidentally, do not look for the spot yourself, because it requires a very precisely made disk, and a precise optical alignment and distance to the screen. Prior to Arago's observation the particle theory had been more accepted than the opposing wave theory of light that originated with Christiaan Huygens. Probably the great prestige of Isaac Newton who developed the particle theory was a major factor in promoting support for it. A possible parallel may exist between Newton's prestige in promoting a dim view of the wave theory and Einstein's reputation in fostering a rejection of FTL tachyons by many of today's physicists.

When my 3 + 3 model was first proposed, I understood that it had little chance of being accepted, given its exotic nature. However, my degree of belief in the model increased as I found each new piece of empirical evidence supporting it, and no definitive evidence against it. As the evidence started piling up, I felt like a kid who had found a large gleaming gold nugget that looked more and more like the real thing. Of course, I remained aware that until KATRIN's results came out, the nugget could still turn out to be the physicist's equivalent of fool's gold, a substance that has deceived many would-be prospectors. Nevertheless, when I discovered evidence for an absurd 8 MeV neutrino line in the SN 1987A data, I think I understood how Arago must have felt in discovering a bright spot in the middle of the darkness. Obviously, I cannot claim that the situation with my 3 + 3 model is completely analogous to the wave-particle debate because the status of the 8 MeV neutrino line and the other two unicorns of Chapter 3 is currently uncertain. However, regardless of how the story finally turns out, the two situations are parallel in that finding a weird predicted empirical effect establishes the validity of the original model or theory that predicted it.

A Third Approach to Physics Research

Most research in physics fits neatly into one of the two categories: theory and experiment. Today some physicists might undertake both experiment and theory, but probably only a negligible number can do both with great proficiency, as Enrico Fermi once did, given the highly specialized nature of the skills required for each activity. This chapter focuses on a new research method of making discoveries in physics called data prospecting, with examples taken from the history of physics, as well as my two-decade long hunt for evidence that some neutrinos are tachyons. The data prospecting method is a useful method, even if it should turn out that "my bright spot in the darkness" is just a defect on my retina.

As previously noted, the data prospecting approach does not fit neatly into either category of theory or experiment. Data prospectors look for anomalies in published data in order to formulate and then validate an unconventional hypothesis. This method differs from the better-known method of data mining, so to clarify the difference let us recall the distinction between the words *"mining"* and *"prospecting"* in connection with mineral resources. *Prospecting* is the act of looking for a spot where there may be valuable ore, whereas *mining is* the extraction of the ore from such a place. The best prospecting sometimes occurs in previously abandoned mines. Thus, prospecting is a more appropriate term than mining to describe the act of extracting evidence from previously published data, which is akin to "abandoned mines."

In contrast to prospecting, data mining is the process of discovering patterns in large datasets utilizing methods at the intersection of machine learning, statistics, and database systems. Thus, unlike prospecting, data mining implies the use of both specific tools and massive datasets, such as those involved in fields such as particle physics and astronomy. In particle physics, for example, experimenters often find themselves inundated by data. Some datasets in this field often consist of exabytes (10^{18} bytes or a billion gigabytes!) of information. The enormous size of an exabyte may be better appreciated if one realizes an exabyte is 20,000 times the amount of information in all the books ever printed! Processing such volumes of data clearly requires cutting edge computers and massive amounts of storage – see Figure 6.2. It is no accident that people in the field of particle physics have been in the

FIGURE 6.2
The supercomputer at Fermi National Lab that has allowed particle physicists to manage and process enormous amounts of data.

Source: Reidar Hahn/Fermilab

forefront of making computational advances, given the demands of their field. In fact, Tim Berners-Lee, a British scientist, invented the World Wide Web in 1989, while working at CERN. He conceived of the Web as a vehicle to meet the demand for automated information-sharing between scientists in universities and institutes around the world.

The field of data mining has many specific algorithms and tools, such as decision trees and neural networks, for identifying rules to classify subsets of the dataset. Clearly, prospecting differs from mining because unlike mining, it can be used with datasets of any size and uses no specific tools. In both data mining and prospecting, the goal is to identify patterns in datasets without any need to specify a hypothesis in advance, but prospectors, unlike miners, explicitly seek to generate interesting and unconventional hypotheses, which can then be tested using other datasets. Crucially, prospectors must not be constrained by the interpretation of the data offered by the experimental group that produced it. Moreover, they always must inspect the *original* data and not just reports about it. Thus, for example, the 3 + 3 model inferred from the SN 1987A neutrino data was developed by ignoring the usual assumption that the neutrino masses needed to be very close together (as required by the standard neutrino model). In addition, support for the model hinged on ignoring the claim by the Kamiokande group that their data for 997 events for the hours before and after the main burst of 12 neutrinos were only due to background.

When developing a model based on an alternative interpretation of data, the wise prospector must have a flexible attitude toward the limits of the possible. There is a famous saying attributed by author Arthur Conan Doyle to his fictional detective Sherlock Holmes: "Once you eliminate the impossible, whatever remains, no matter how improbable, must be the truth."[2] Holmes' aphorism may wrongly suggest that it is easy to identify and eliminate what is impossible in one's pursuit of the truth. Of course, many things known to be possible today would have been dismissed as impossible in an earlier era, including for example, teleportation, quantum entanglement, and light behaving paradoxically like both a wave and a particle.

One might argue that we can only rule out as impossible that which violates the known laws of physics, but even that formulation may be too restrictive, given the provisional nature of physical law. Recall that prior to Wolfgang Pauli's hypothesis of the neutrino, consideration was even given to abandoning the conservation of energy law, except on a statistical basis. Even though that approach proved wrong, the possibility was evidently worth considering. Therefore, in seeking to develop an unconventional model do not be too sure that you know what is impossible, lest you eliminate something that is absurd but true – recall the example of Arago's spot! In contrast to the fictional Sherlock Holmes approach, I much prefer that of science fiction writer Arthur C. Clarke, who noted: "The only way of discovering the limits of the possible is to venture a little way past them into the impossible."[3]

In the history of physics, an example of someone who was *almost* a data prospector is Max Planck. In 1900 he observed the spectrum of radiation from a so-called blackbody (that is, an object that absorbs all wavelengths of radiation incident on it). Planck found that the observed spectrum of a heated blackbody disagreed with the theoretical predictions (a conflict known as the "ultraviolet catastrophe"). He was able to resolve the conflict by making two *ad hoc* assumptions: (1) that the energy is emitted in "chunks" or quanta, and (2) that the energy of each quantum is proportional to the light frequency. (These quanta are now called photons.) Planck's assumptions lacked any theoretical foundations, and in fact are incompatible with classical physics. Writing three decades later, Planck noted that he regarded his idea of light quanta as "a purely formal assumption," and wrote "I really did not give it much thought except that no matter what the cost, I must bring about a positive result."[4] By the phrase "positive result," Planck was admitting that he had resorted to "curve fitting," that is finding the mathematical function that fits the data best, and he hoped that someone would eventually discover the real explanation of the blackbody spectrum's shape.

Had Planck truly been a data prospector, he would have taken his outlandish assumption of light being emitted in quanta seriously, and then considered what other kinds of data (such as that arising from the photoelectric effect) that his assumption might also fit. Even after Einstein took that very step in 1905, Planck still considered the assumption that light energy comes in quanta simply a mathematical convenience that did not refer to actual energy exchanges between matter and radiation. Only in 1908 did Planck convert to the view that the quantum in his formula for the blackbody spectrum represented a new reality beyond the understanding of classical physics. Interestingly, Einstein was awarded the Nobel Prize not for relativity, but for the explanation of the photoelectric effect in terms of light quanta.

Pros and Cons of Data Prospecting

Particle physics experiments, because of the immense datasets they generate, may offer opportunities to data prospectors, as a promising way to uncover some important results. This is especially true for those who are solo researchers – or as I prefer to call them "lone wolves" rather than "pack hunters," which might describe the hundreds or sometimes thousands of physicists usually involved in such experiments. Data prospecting is especially appealing to someone lacking both the deep understanding of mathematics needed to be a good theorist and the abilities of a skilled experimentalist, except those connected with data analysis. Of course, being a member of one of the large teams that discovers a new particle like the Higgs also has a certain allure. However, you will be just one cog in a very big machine, even if it may inflate your publication record. The prospecting approach certainly is not for everyone and is best suited to those seeking to make a major discovery on their own.

The downside of adopting the data-prospecting method is that often when you observe an anomaly or a pattern in published data, nothing useful or important is discovered. Furthermore, doing data prospecting may make you appear as a crackpot in some eyes, especially if the work is done on a highly controversial topic like tachyons. On the positive side, when work is done in an unconventional area that few take seriously, few competitors exist.

Pursuing work on a highly fashionable topic is not unlike choosing to climb Mount Everest, which would likely place you in a very large crowd of other researchers – see Figure 6.3. According to physicist and inventor Luis Alvarez the worthiness of a research topic equals the product of the likelihood of success and the significance of a successful result. Thus, unconventional topics on the fringes of the possible may be very worthwhile even if the odds of success are small – provided that a successful outcome would lead to a major advance. You must, however, be mindful of the low chance of success in such a pursuit, making such topics a high-risk, high-reward activity.

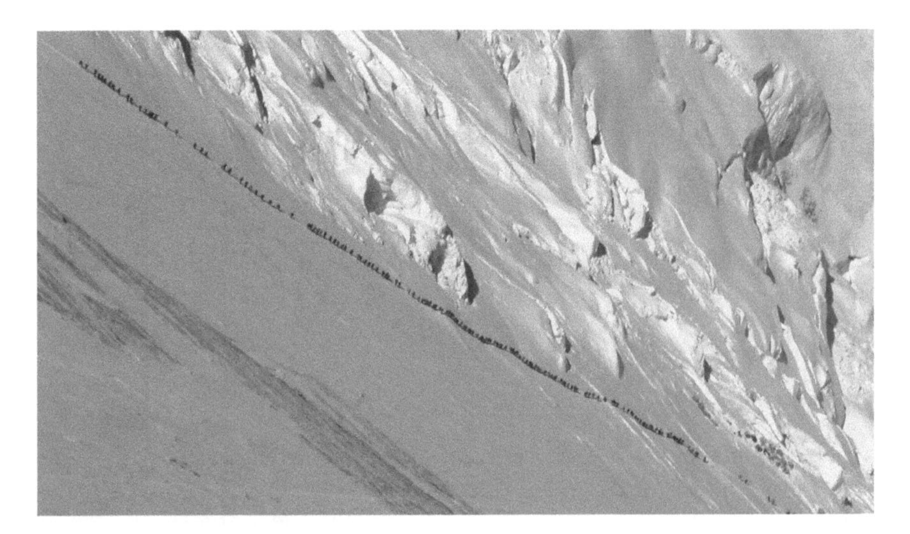

FIGURE 6.3
A long line of climbers moving up Mt Everest in 2012.

Source: Photo by Ralf Dujmovits, professional climber

Negative Evidence and Reviews

When doing data prospecting you must diligently seek evidence that can disprove your model, as well as positive evidence to support it. Seeking negative evidence is even more important, because if you find it you will know that you should move on and work on more productive pursuits. The most important type of negative evidence is data that appear to refute your model. However, when negative evidence is found sometimes it can be analyzed in such a way as to be consistent with your model. In these cases, such "failed refutations" which do not kill your model only make it stronger!

An example of how negative evidence can be interpreted to support a model that it seemingly contradicts was discussed in Chapter 5. The initial results from a limited amount of KATRIN data for tritium beta decay were consistent with a single neutrino effective mass less than 1 eV, but seemingly not with the 3 + 3 model. This apparent contradiction between the first KATRIN data and the 3 + 3 model was resolved by simply assigning to each of the three masses in the model, different contributions of each mass to the overall spectrum. My earlier claim about pre-KATRIN experiments favoring the 3 + 3 model had been a mistake – the earlier experiments just were not sensitive enough to distinguish the model from the standard one. Thus, with a different weighting of the spectral contributions for the three model masses, the earlier experiments and KATRIN's first results were both consistent with either a single small effective mass, or the 3 + 3 model.

Negative evidence sometimes will come from a referee who points out flaws in a paper that you might have submitted prematurely. Mistakes are far easier to make by lone-wolf data prospectors than pack-hunters. No one likes negative reviews, but even when reviews are ill-informed (or in some cases even hostile) one must ignore nasty comments, look for the legitimate ones, and try to deal with them – either by modifying the model or your description of it. I have an extensive familiarity with negative reviews, having published 18 tachyon papers in peer-reviewed journals over the years, of which all were initially rejected, many more than once.

Anyone publishing in a controversial area such as tachyons should expect negative reviews. In fact, the very first paper on tachyons by George Sudarshan and his colleagues was rejected when it was submitted to *the Physical Review*. A revised version was later submitted to a less prestigious instructional journal, where after some "informal lobbying," it was finally accepted. Moreover, given its submission to an instructional journal, the article could be framed as more of an "exercise for students," rather than a serious research proposal.

The most frustrating of my negative reviews were from referees who simply ruled out the possibility of tachyons on theoretical grounds as "unphysical," without even bothering to rebut the evidence presented. One brief negative review noted that "Ehrlich sees tachyons wherever he looks." I should have replied to this comment by noting that this was because I looked for tachyons exactly where evidence for them would be found if they really exist. Nevertheless, despite many negative reviews over the years, I am truly grateful for all of them (even the nasty ones), because they encouraged me to develop my ideas more fully, and they prevented some very seriously flawed early versions of my papers from getting published!

Fake or Predatory Journals

It is important to learn from negative reviews and not be discouraged by them. Researchers in a controversial area should not yield to the temptation to publish in journals having zero or minimal peer review. These publications, often called fake or predatory journals have proliferated enormously in recent years. Fake journals offer authors a financially costly and unethical way to inflate one's credentials by publishing papers that would not be accepted by legitimate journals. Rather than the expanding human knowledge, the main motivation for the existence of such journals is profit for their creators. Submissions of computer-generated complete rubbish have been accepted by some fake journals, usually for a sizable fee.

Many signs point to a fake or predatory journal; one is that they have names very similar to existing well-known journals, or they solicit paper submissions, often through a flattering email. Such solicitations often tout the journal's "impact factor," and promise quick publication, but they may initially neglect to mention that some significant payment will be required. Of course, not all journals that require a publication fee fall in the predatory category, since a fee is a standard practice for most "open access" journals, which can be read and downloaded by anyone at no charge.

Predatory journals sometimes do claim to be peer-reviewed, but their reviews are often a sham. Another tell-tale sign of a fake journal is that the email soliciting your submission is grammatically incorrect. In other cases, the persons on the editorial board are either unnamed, or the list arouses suspicion. For example, one publishing conglomerate that has been listed as being predatory publishes 222 different journals and claims to have nine Nobel Laureates on its various editorial boards.

When fake journals do have reputable people listed as the editors, sometimes the listing is done without their permission. In one case, a reputable scholar accepted an invitation to become chief editor, and then found that he had no control whatsoever over "his" journal, and subsequently discovered to his dismay that he was unable to renounce this bogus editorship. Websites exist that list predatory journals, but new ones pop up so fast that it is difficult to keep up with them. Moreover, some people who have maintained lists of such journals on their website have faced legal action. Solicitations of manuscripts in fake journals often come from fields well outside your own. Scholars who are most susceptible to being duped by fake journals are usually at the start of their research careers and may not be familiar with the reputable journals in their field. The most vulnerable researchers are probably from developing nations and may be less able to spot the flawed grammar or exaggerated prose in an email soliciting their submissions. Finally, those researchers from developing nations might be overly impressed by journals having the word *American* in their titles, or those which appear to have American editors.

Aside from cheating unsuspecting scholars who are charged for submitting papers (that are sometimes never published), these fake journals, and many fake conferences, are a cancer on the research enterprise. Over time their existence could blur the distinction between real and fake research, especially if increasing numbers of scholars with such fraudulent accomplishments on their resumes find their way into the ranks of academia. Academics who have difficulty publishing will likely regret publishing in fake journals, unless perhaps their academic department already consists of fakers. Perhaps just as bad as publishing in fake journals is self-publishing research that was rejected by a legitimate journal. Even worse is sending your self-published theories unsolicited to a wide range of people who are unlikely to pay any attention to it.

Making as Many Mistakes as You Can

If you are trying to find evidence for a new model or theory, inevitably you will make mistakes. Although mistakes are common in research, they are especially likely if you are working without a collaborator, which is often the norm for those pursuing an unconventional model. For the data-prospector seeking validation of an unconventional model, a trusted knowledgeable confidant who you can ask for critical comments on your work before you submit it for publication can be extremely useful.

Keep in mind that even the greats, including Einstein, have on occasion made major mistakes before their ultimate triumphs. While he was developing general relativity, Einstein had proposed a test of his theory based on the slight deviations in the apparent position of stars located along directions near the sun due to the deflection of starlight by the sun's gravity. These deviations, of course, could only be observed during a solar eclipse when some stars near the sun can be seen. However, in making his prediction Einstein

FIGURE 6.4
1919 Solar eclipse observed by Dyson, Eddington, and Davidson. The image is the result of modern image processing techniques. Six faint stars have been labeled on the image.

Source: Photo from ESO Observatory

had initially forgotten to account for the bending of a light ray due to the warping of time as well as space. As a result of this omission, his prediction was off by a factor of two. World War I was a great tragedy for millions of people, and Einstein, a devout pacifist, supported anti-war movements all during its duration, which makes it ironic that the war might have saved his career.

An English expedition to measure the solar eclipse of 1914 and verify Einstein's prediction found itself stranded in Germany at the outbreak of the Great War, unable to perform the crucial test of his then incorrect prediction. Fortunately, Einstein found his mistake and corrected it before the eclipse observation was made after the war in 1919 (see Figure 6.4). Imagine what might have happened, however, had he not been so lucky. Would Einstein's doubling of his predicted value after the measurement had been made be interpreted as his throwing in a "fudge factor" to get agreement with experiment?

Often mistakes are valuable because they help us understand what needs improvement or revision – see Figure 6.5. In my own case, I have made many mistakes over the years, but some stand out regarding the 3 + 3 model. I already noted, for example, that I dismissed the idea of an 8 MeV neutrino line from supernovae (needed to explain the Mont Blanc neutrinos as tachyons), as something I found "inconceivable," only to then *conceive* of a way of creating such a line just a few years later. Apparently, I had over-relied on

FIGURE 6.5
"Mistakes show what needs improving, so let's make as many as possible."

Source: Shutterstock image

the assurances of modelers that the spectrum of neutrinos from supernovae could not contain a neutrino line. I also was initially too quick to accept the conventional wisdom that the 5-hour early Mont Blanc burst was just some random event unassociated with SN 1987A.

As another example, I had initially held the mistaken belief that even though the masses contributing to the electron neutrino were very far apart in the 3 + 3 model, the beta decay spectrum still could be described by a single "effective" mass, rather than the combination of three spectra – one for each mass. In fact, I even submitted a grossly incorrect paper trying to justify this misguided belief, which was fortunately rejected. I should have followed the advice of two confidants, who assured me I was mistaken after the flawed paper was written! Probably the two most serious mistakes one can make involve deciding when to stick with a model and when to abandon it. Sticking with a model that negative evidence clearly shows to be unfeasible, is just as bad as abandoning it prematurely, if one can find a way to circumvent negative evidence.

My initial foray in hunting tachyons started over two decades ago, when I found some possible evidence for them based on cosmic ray data. This early work led to several published results, but they could not be corroborated, given the many alternative explanations of cosmic ray data. Regrettably, it did not occur to me at the time to seek further evidence of neutrinos being tachyons in other areas of physics. In other words, I did not do enough "what about...?" type of thinking about the problem. As a result, a decade-long gap occurred in my tachyon hunting activity after my cosmic ray work. My interest in the problem was rekindled only because of the 2011 report by the OPERA group at CERN suggesting that neutrinos appeared to be superluminal – a finding that turned out to be erroneous. Nevertheless, I sincerely thank the OPERA group for their wonderful FTL neutrino anomaly because it reawakened my interest in the subject, and shortly afterwards I wrote a paper proposing the 3 + 3 model.

The decade-long gap in my tachyon hunting activity prior to OPERA, while delaying my progress, also had a positive side effect. The gap allowed me to take a fresh look at what I had done earlier and made me realize that the previous evidence for tachyonic neutrinos was not as strong as I had initially thought. When constantly dealing with skeptical reviewers, it is easy to become overly defensive about your ideas, and to ignore some problematic aspects. The fresh look after a decade made me understand that I needed a wider array of evidence from many different areas to convince others of the validity of my controversial model. That decade-long gap also gave me the opportunity to reach out to the leaders of the KATRIN experiment at just the right time, close to the start of data collection.

During one of two seminars presented to the KATRIN group I ended my talk with a joke or rather a pun. Although I do not know German, I concluded with: "Wenn sie das experiment fehlerfrei durchführen, finden sie die Ehrlich-werte für die neutrinomassen." Approximate translation: "If you do the experiment without error you will find the Ehrlich values for the neutrino masses," but the phrase "Ehrlich values" also could mean "honest values" since my name in German means honest. Anyway, the remark did get a laugh, so I guess my memorized German was understood.

Spotting Promising Anomalies

There are no hard and fast rules for identifying those anomalies in published data which might be especially promising for developing an unconventional model, but they usually share some characteristics. An obvious one is that the anomaly should have high statistical significance, so its chance occurrence is unlikely. In addition, the anomaly should have no plausible explanation within the current paradigm. Furthermore, it is helpful if the anomaly can be looked for in many different kinds of datasets.

Normally, data for a one-time occurrence like SN 1987A might not seem a promising place to find anomalies, particularly given the smallness of the dataset. With only 30 neutrinos in four detectors the statistical significance of any conclusions cannot be very high. On the other hand, the additional data (997 events) that the Kamiokande group had fortunately included in their 1988 paper on SN 1987A – data taken over the hours before and after the short burst – turned out to provide a key result. Without this fortunate inclusion to their paper no well-camouflaged 8 MeV line atop the background (one of the three hidden unicorns in Chapter 3) would have been identified.

Some readers may have in mind the possibility of searching for anomalies in data outside the bounds of physics. The study of anomalies generally is termed *anomalistics*, defined as using scientific methods to evaluate anomalies or phenomena that fall outside current understanding, to find a rational explanation. This term was originally coined in 1973 by Roger W. Wescott, an anthropologist. Wescott defined anomalistics as the "serious and systematic study of all phenomena that fail to fit the picture of reality provided for us by common sense or by the established sciences."[5]

Wescott's definition might suggest that the explanation of interesting anomalies must be found *outside* the established sciences. Anomalistics by his definition therefore includes such pseudo-sciences as ufology, and parapsychology, with which many legitimate scientists do not wish to associate. People who work in such areas often submit their findings to fake journals or conferences, which are not run by mainstream scientific organizations, but rather societies that are committed to studying pseudoscience topics. The motivation for avoiding mainstream journals by people working on pseudoscience topics is probably rooted in their belief that the established sciences are too closed-minded to consider such work impartially. While this concern may have some validity, a bigger problem is the gullibility and the low standard of evidence adopted by people working in these areas. I once attended a conference on the study of anomalistics, and I observed that virtually all speakers were greeted with an affirming uncritical attitude, even when their claims were not very well-supported, to put it mildly. This type of mutual support, while it may engender warm feelings, was quite unlike what I have observed at physics conferences, where some presentations were challenged severely if the speaker did not note that the supporting evidence was weak.

Search for Extraterrestrial Intelligence (SETI)

Fringe science topics even "ufology," need not be beyond the pale if treated properly. One good example of data-prospecting leading to an interesting highly controversial hypothesis in the ufology area is *Oumuamua*, the first ever observed interstellar asteroid in 2017. Abraham Loeb, former chair of the astronomy department at Harvard University and his associate Shmuel Bialy, had puzzled over some anomalies in the published data on *Oumuamua*.

FIGURE 6.6
Artist's concept of the conventional view of *Oumuamua* as it passed through the solar system.

Source: ESO Observatory/ M. Kornmesser

Such objects, by definition, are not orbiting the sun, so they follow a hyperbolic path (not an ellipse) and they have much higher speed than a normal asteroid in a solar orbit. Loeb and Bialy wrote a 2019 paper identifying six characteristics that taken together suggest *Oumuamua* may have an artificial origin. For one, Loeb claims that *Oumuamua* is moving too fast to be an inert rock, and its motion is as if it is being propelled by a rocket engine. Another surprising aspect of *Oumuamua* is that if it is simply an inert rock, it never should have been found. In other words, for it to have been observed, each star in the Milky Way would need to eject more than a hundred times as many such objects during its lifetime as we expect, based on models of our solar system.

Loeb further argues that *Oumuamua* probably is not a clump of rock shaped like a long potato, as most astronomers believe – see Figure 6.6. Instead, he maintains that its behavior during the 34 days of observation is explained better if it is shaped like a 1 km diameter pancake at most 1 millimeter thick, having a very shiny surface. The preceding description is what one might expect for a solar sail, which would utilize the pressure of light to propel an interstellar vessel. In the case of one sent from Earth, the light might be supplied by a large bank of lasers aimed at the sail driving it forward. Loeb and Bialy's hypothesis is highly controversial among astronomers, and they are likely to be mistaken. However, their proposal is a nice example of data prospecting, since they used published data on the asteroid to draw a highly unconventional conclusion. Unfortunately, there are no other datasets yet that might confirm their hypothesis. So, we may need to wait until the next "visitor" arrives, since *Oumuamua* is now too far from Earth to make further observations. There now has been a second "visitor" to our solar system known as 2I/Borisov, but it appears to be an interstellar run-of-the-mill comet, which stands in stark contrast to *Oumuamua*.

Aside from the reputation of the person making an unconventional proposal (which is "stellar" in Harvard astronomer Loeb's case) the other key indicator in whether fringe topics such as intelligent aliens are being treated seriously is where the results are disseminated. Loeb's article about *Oumuamua,* for example, was published in a top scientific journal (*Astrophysical Journal Letters*), and not in journals devoted to claims of paranormal phenomena.

THE CURIOUS CASE OF JOHN MACK

There is a certain instant credibility when a subject like UFO's is taken seriously by a Harvard academic, but of course, even Harvard academics can be mistaken or even deluded. In the early 1990s, Harvard Psychiatry Professor and Pulitzer Prize Recipient John Mack studied 200 men and women who reported recurrent alien encounter experiences. Mack initially suspected that these people were suffering from mental illness, but after interviewing them he concluded these individuals were quite normal, and they were not fabricating their claimed experiences.

During the 1990s, the Dean of the Harvard Medical School appointed a committee to review Mack's clinical investigation of the people who had shared their alien encounters with him. The committee's draft report suggested that "To communicate, in any way whatsoever, to a person who has reported a 'close encounter' with an extraterrestrial life form that this experience might well have been real ... is professionally irresponsible."[5] The final report removed that statement following a massive legal battle and it merely censured Mack for what were considered methodological errors. Moreover, following release of the final report, the Dean concluded "Dr. Mack remains a member in good standing of the Harvard Faculty of Medicine,"[6] and he took no action against him.

In case you're wondering about the nature of Mack's methodological errors, it was his use of hypnotic recall of their abduction memories. The weight of the evidence about hypnotic recall has shown that it gives you what you expect to find. In other words, if a psychiatrist expects to find memories of alien abduction, childhood sexual abuse, or past lives, that's what he or she will find. In fact, in 1985 the American Medical Association took a stand warning against accuracy of memories recovered through hypnosis. Of course, what really troubled the committee was mainly not Mack's belief in recovered memories under hypnosis, but rather the outlandish use to which it was applied. Nevertheless, the legal pressure brought to bear on the committee probably led it to settle on the less severe charge against Mack, and the subsequent lack of action by the dean. Unlike John Mack, Avi Loeb appears to have no such skeletons (or aliens) in his closet, and he is treating the search for ET in a sound manner using conventional scientific methods.

Loeb's most important contribution to SETI, however, probably will not be his very speculative interpretation of *Oumuamua* as an artificial object produced by an extraterrestrial civilization. Rather, it will probably be his launching of the Galileo Project. This project begun in 2021 involves the systematic search for evidence of extraterrestrial technological artifacts. The history of astronomy has shown that new types of telescopes have opened entirely new ways to see the universe – not just in visible wavelengths, but also in radio, infrared, X-ray, and gamma rays. In the same manner, Loeb hopes that new types of high-resolution telescopes will dramatically increase our ability to look for alien artifacts. The most likely kind of artifacts would probably not be large scout ships of *Oumuamua's* size, but rather they could be very small "message-in-a-bottle" vehicles. Recall that NASA has already launched two such small Voyager interstellar probes into deep space 44 years ago, carrying greetings from Earth to anyone who receives it. Loeb's Galileo Project plans to make publicly available all the data it gathers. These data would include both future *Oumuamua* type objects, as well as much smaller objects akin to the voyager vehicles that some alien civilization might have launched.

The subject of SETI might be a very fruitful area for data prospecting. Aside from future data gathered by the Galileo Project, there already exists publicly available data recorded at various frequencies from many possible stellar sources. The lack of any identification of an artificial origin to date need not imply the absence of such a signal, simply the absence of an identifiable one – perhaps we just did not know what to look for? Do we even have a clue, for example, what some highly intelligent creatures, such as whales on planet Earth, are communicating to one another (or maybe to us) with their songs? Some have even suggested that the surest sign that intelligent life exists elsewhere is that they have *not* tried to contact us!

The best known existing open data repository of SETI data was created by a group at the University of California at Berkeley under the "Breakthrough Listen" project, funded by Internet billionaire Yuri Milner. The project announced in 2015 will generate as much data in one day as previous SETI projects generated in one year. All the data can be downloaded by anyone, and all interested parties have been invited to help develop software and algorithms. Targets for the project include one million nearby stars and the centers of 100 galaxies, with observations in both radio and optical wavelengths. Although a million nearby stars represents an impressive advance over previous efforts, it represents a mere 0.0005% of the stars in our galaxy, so if there were only on average one civilization per galaxy, the chances are slim we would find anyone out there.

The first candidate for a possible SETI signal called BLC1 (Breakthrough Listen Candidate 1) was announced in December 2020. The signal appears to emanate from the direction of our neighboring star, Proxima Centauri, and it cannot yet be dismissed as merely Earth-based interference. The radio signal has a precise frequency of 982.002 MHz, which would not be produced by any known natural process and was detected during 30 hours of observations in 2019 by the Parkes Observatory in Australia. However, as of December 2021, follow-up observations have failed to detect the signal again, which is one self-imposed test necessary to confirm that it was a true SETI signal.

Dyson Spheres

Another SETI approach that some researchers have considered involves looking for so-called *Dyson spheres*, named for physicist Freeman Dyson. These objects are hypothetical megastructures that an advanced civilization might construct to extract energy from the star that their planet orbits – see Figure 6.7. Dyson in a 1960 paper noted that once an advanced civilization exhausted the energy resources of their home planet it might seek energy by building megastructures consisting of solar energy collectors surrounding its sun. These Dyson spheres would enable the aliens to tap into vastly greater energy resources. For example, if our sun's full output could be harnessed, it would provide roughly 20 trillion times the power currently used by humanity. Dyson noted that such spheres might produce specific observable signatures that would allow their detection – specifically an enhanced amount of infrared radiation coming from the star they surround.

FIGURE 6.7
Dyson sphere around a distant star in front of the Milky Way.

Source: Image provided by Shutterstock

While Dyson was the first to write a scientific paper on the subject, he credited science fiction writer Olaf Stapledon with the idea. Dyson even suggested, in a commendable gesture of self-effacing humility, that the hypothetical objects be known as *Stapledon Spheres.* In his 1937 novel *Star Maker* Stapledon had described "every solar system ... surrounded by a gauze of light traps, which focused the escaping solar energy for intelligent use."[7] As was the case in the novel, these spheres are normally not considered to comprise a solid spherical shell of matter surrounding a star, which would likely be impossible to construct out of any known material. Rather, the sphere is assumed to consist of either a mesh or a swarm of orbiting satellites, both of which would absorb only a small fraction of the sun's energy output. The orbiting satellite design would have the advantages of not presenting severe stability challenges and since modular, could be built in stages. Although Dyson spheres could not be built with current technology, futurist George Dvorsky has argued that the technology for constructing such a megastructure need not be in the far distant future if self-replicating robots were used in the construction. No Dyson spheres have yet been found, but astronomers have searched for them by examining the infrared spectra of various stars. If a star were surrounded by a Dyson sphere, we would expect to see a spectrum consisting of a sum of two blackbody radiation curves at very different temperatures. One such search examining a million stars and turned up 17 potential (but ambiguous) Dyson sphere candidates, of which four were considered still questionable but only "slightly amusing." Apparently, astronomers are not easily amused.

Are We Alone?

Obviously, SETI would be a more promising endeavor if there were a very good chance we are not alone – see Figure 6.8. Quite a bit of guesswork is involved in estimating how many intelligent and technological alien civilizations exist out there. Most researchers have been relatively optimistic, however, with 64% estimating over 100 of them in our galaxy alone. On the other hand, very pessimistic estimates also exist. One recent study has suggested that the odds that we are currently the only intelligent and technological life in the Milky Way Galaxy range between 53 and 99.6 %, so based on that study we could well be alone in the galaxy at present. The phrase "at present" here suggests the dismal notion that many other intelligent civilizations may have existed, but they pop up briefly from time to time only to quickly disappear, because of their suicidal behavior.

Most estimates of the number of *communicative* intelligent civilizations make use of an equation originally written by astronomer Frank Drake. The word *communicative* here implies that the civilization would reveal their existence via signals they release into space. There are many possible reasons why a civilization might choose not to try to communicate with

FIGURE 6.8
The world's first ant colony to achieve sentience calls off the search for us.

Source: Courtesy of xkcd comics (xkcd.com)

others, but the word communicative here includes both deliberate and non-deliberate emissions of signals. You could easily use the Drake Equation to make your own estimate of the number of alien civilizations using an online calculator by entering your best guess as to the values of each of the seven parameters in the equation. Given the large uncertainties in each parameter, your guess may be as good as that of some experts. In fact, when Drake first wrote his equation in 1961, he recognized that it was merely a way to stimulate scientific dialogue about SETI, and not yield a precise estimate.

The most uncertain of the seven factors in the Drake Equation is the average longevity of an intelligent species once it develops technology capable of destroying itself. The authors of the very pessimistic study estimate this factor is anywhere between 100 years and ten billion years. Thus, in a manner of speaking, the probability of identifying intelligent extraterrestrials capable of sending signals to us is closely related to whether you believe there are any species of intelligent *Earthlings*. In particular, the probability depends on whether we are smart enough not to destroy ourselves relatively soon now that we have acquired the technology to do so.

Our Chances of Making It to a "Post-Human" Era

Given the seriousness of the problems now facing us, and the short-term thinking of most of humanity, pessimism regarding the prospects of our very long-term survival is understandable. While I do not consider the chances of our long-term survival to be very great, neither do I think they are negligible. My guess, based on nothing more than intuition, of our chances of making it through the coming turbulent decades (probably the hard part) and then evolving to a "post-human" civilization that might last millions or billions of years is perhaps at most 10%.

One way of finding out the likelihood that advanced ET civilizations destroy themselves would be to look for evidence that such self-destruction has taken place either on Earth or other planets. If there had been a previous technological civilization on Earth, any manufactured artifacts would only survive over timescales of a few thousands of years.

Even for civilizations existing a few thousand years ago, however, we may know very little. The discovery of the Antikythera mechanism, named after a Greek island near which it was found in 1901, serves as a reminder of how ignorant we have been about civilizations that existed not that long ago. This piece of ancient Greek technology was arguably the world's first computer – see Figure 6.9. Historian Alexander Jones has characterized the discovery as "Like opening a pyramid and finding an atomic bomb."[8] In 2008, a team led by Mike Edmunds and Tony Freeth at Cardiff University used X-ray tomography to image inside fragments of the crust-encased mechanism and read faint inscriptions that once covered the casing of the machine. These inscriptions suggest it had 37 meshing bronze gears enabling it to follow the movements of the Moon and the Sun through the zodiac and to predict eclipses.

Going back say a few million years, everything will have turned to dust. We, therefore, cannot dismiss the possibility of a previous pre-human technological society existing (and then going extinct) over very long timescales. In fact, Adam Frank, an astrophysicist at the University of Rochester, and climatologist Gavin Schmidt at NASA, concluded in a 2018 paper that only

FIGURE 6.9
The Antikythera mechanism for tracking the sun and the five then-known planets. Left: One of several fragments that was found, right: recreated mechanism © 2020 Tony Freeth

very indirect evidence would exist for a pre-human civilization. The idea that such a civilization may have existed millions of years ago is known as the *Silurian Hypothesis*, a name inspired by the fictional Silurians in the Doctor Who TV series. Incidentally, this possibility, while it is quite speculative, has been given some credence by astrophysicists Iosif Shklovsky and Carl Sagan in their classic book *Intelligent Life in the Universe* (1966).[9] The Silurian hypothesis, however, should not be confused with claims of "ancient astronauts," which are not taken seriously by most scientists, who consider it to be pseudoscience. The believers in ancient astronauts typically claim without much evidence that humans are either descendants or creations of extraterrestrial intelligence who landed on Earth mere thousands of years ago.

One might also seek evidence of extinct civilizations on other planets, especially the many planets now known to orbit stars other than our sun. Seeking evidence for advanced ET's self-destruction, three astrobiologists, Adam Stevens, Duncan Forgan, and Jack O'Malley James have considered various scenarios and what the observable signatures might be in each case. Although only a small fraction of their proposed self-destruction scenarios might be detectable over interstellar distances, a few might be detectable with next-generation telescopes. The most easily detectable possibility would involve finding significant changes in a planet's atmospheric composition, say following a nuclear war or a runaway greenhouse warming. However, in the case of a runaway greenhouse warming in the case of a planet around another star, it would be near-impossible to show it was not due to natural causes – as probably occurred with the planet Venus.

If the number of observed self-destructions out there should be significantly greater than the number of ET's we observe (currently zero), such a finding would not bode well for our own chances of avoiding self-destruction. Recall my earlier "guestimate" that our chances of evolving to a "post-human" civilization that might last millions or billions of years is perhaps at best 10%. If this guess has any relation to reality, then a very interesting consequence follows, according to a Swedish philosopher Nick Bostrom. A professor at the University of Oxford, Bostrom has written a provocative 2001 paper, in which he asks the question posed by the title of the next section.

Are You Living in a Computer Simulation?

In my ninth decade of life, as I merrily row, row, row my boat down the "stream of time," I find that I very much agree with my late father's often-expressed notion that life is but a dream. My father, who graduated from college with a

FIGURE 6.10
The author with his parents near the beginning of his "dream" outside their home in Brooklyn, New York

degree in engineering during the depths of the Great Depression was unable to pursue that field, so he worked as a postal clerk for most of his career to support his family. Regrettably, I did not always appreciate the wisdom of my Dad, who I always called "Bill," until my adulthood. That anxious-looking little kid in Figure 6.10 is me between my parents. The photo was taken outside our Brooklyn, New York home around 1940, just a few years after my dream began.

The only tweak I would make to my dad's belief that life is but a dream is that I suspect our dream is in fact a super-sophisticated computer simulation, much like that in *The Matrix* movies. Even with today's computers we have simulations that allow users to interact with their 3D virtual environment and each other in a highly realistic manner, in which virtual reality becomes almost indistinguishable from "real reality." If the exponential advances in computer technology were to continue unabated, it is likely that an advanced post-human civilization living many centuries from now would have unimaginable amounts of computing power.

Suppose we assume that our post-human descendants will have an interest in finding out about their ancestors, and they have the ability and interest in running extremely realistic computer simulations of any past historical era. Philosopher Bostrom, whose famous paper was used for the title of this section, has made a radical suggestion about the simulated people in the simulations that vastly more powerful computers would make possible:

> Suppose that these simulated people are conscious ... Then, it could be the case that the vast majority of minds like ours do not belong to the original race but rather to people simulated by the advanced descendants of an original race.[10]

The Singularity is a hypothetical point in time when technological growth becomes irreversible, resulting in unforeseeable changes. In one version of the singularity hypothesis, a "runaway reaction" of self-improvement cycles, with each new and more intelligent generation appearing more and more rapidly and resulting in a powerful superintelligence that far surpasses all human intelligence. Polls of AI researchers, conducted in 2012 and 2013 by philosopher Nick Bostrom put the estimated time of the singularity in the 2040s.

Bostrom has also assumed that consciousness need not only reside in a biological brain, but it could also exist in a computer. Given Bostrom's assumptions, one might imagine that a post-human civilization lasting perhaps millions or even billions of years would run enormous numbers of simulations corresponding to different epochs during which their ancestors lived. Suppose for example that the typical post-human civilization ran ten million simulations of our current era, and that the simulations are so good that conscious beings in one of them do not know if they are in a simulation or in the original world being simulated. If they were mathematically oriented, they might in fact ask: "what are the odds that I am living in a simulation?"

Guessing the Odds

Suppose we assume: (1) only 10% of civilizations will not self-destruct before evolving to become a long-lived post-human civilization, and (2) they might generate ten million simulations for any given era over some extended period. These assumptions would mean that one million simulations are done of our current era compared to the one original occurrence. We would, therefore, find the odds of a person being in a simulation rather than being in the original epoch were 1,000,000 chances out of 1,000,001, or 99.999%, which is a virtual certainty! We might want to distinguish, however, between simulations of a particular era and simulations of a particular person's life in that era. In the latter case, our estimation of the odds would depend on the identity of the person making the estimate. Thus, if the hypothetical person were someone like Lady Gaga or Marie Curie, a great amount about their lives is likely to survive in the historical record. As a result, there could easily be 1000 times as many simulations of their lives done than for the average person, so such well-known people might justifiably be even more certain they are living in a simulation. This observation led me to speculate that there might be many more simulations done for the life of the guy who discovered tachyons than for some delusional schlub who merely imagined he discovered them – but of course this speculation *presumes* that tachyons in fact exist!

The whole idea that we are in a computer simulation may sound crazy, but psychiatrists assure us we need not actually be crazy to believe it. On the other hand, there is a psychiatric condition known as "depersonalization disorder," afflicting 1 or 2% of us, in which you persistently feel that you have a sense that things around you are not real, and you are living in a dream or a computer simulation. Many of us have such feelings occasionally, but they reach the disorder stage only when these feelings interfere with our ability to function. Of course, there remains the possibility that these feelings arise not because of a "disorder," but because we really are in a computer simulation, and those afflicted are the more perceptive among us.

Although Bostrom's thesis that we are likely living in a computer simulation is highly controversial, no one has yet pointed out serious flaws in his reasoning, and it has even been embraced by luminaries including Elon Musk and astrophysicist Neil de Grasse Tyson – or perhaps their simulated descendants! In fact, Musk thinks it is almost certain that we're living in some version of *The Matrix;* he goes so far as to speculate that there's only a "one in billion chance" that reality as we know it is *not* a computer simulation.

Bostrom in his 2001 article expressed his simulation hypothesis in more neutral terms than the preceding summary may have suggested. In the article he merely noted that at least one of the following three things must be true, and he left it to the reader to choose which one(s) they might be:

1. Humanity has a negligible chance to survive to a posthuman stage.
2. Any posthuman civilization is ***un***likely to be interested in running simulations of their history.
3. We are extremely likely to be in a computer simulation.

Given Bostrom's reasoning, the idea that we are beings in a simulation (number 3) would have to be true if propositions 1 and 2 are both false. You may find the notion of your being a conscious being in a simulation too hard to swallow. In that case, you may find it easier to imagine that what you call your life is instead some being or entity experiencing a super-realistic virtual reality. Of course, that being, namely you, would not be just wearing virtual reality goggles, but all your senses are involved in the simulation. Even now, one can experience haptic virtual reality simulations, where with special gloves your hands feel what you see with special goggles.

How Could We Tell?

For the simulation hypothesis to be a promising research topic for a data prospector, a possible way to test it must exist. Some researchers have proposed tests that involve looking for a "glitch in the matrix." The phrase comes from the movie *The Matrix,* where a 'glitch' drew the hero's attention to the fact that he was in a simulation. The specific glitch apparently occurred when new computer code was introduced into the simulation. This caused the hero (Neo), played by Keanu Reeves, to experienced strange déjà vu sequences, which occurred within seconds of each other. Another possible way a glitch might reveal itself is to find ways that the simulation designer may have "cut corners" or made approximations in its creation. For example, one might imagine that the underlying structure of a simulation was performed using a space-time grid of finite size intervals. Three physicists Silas Beane, Zohreh Davoudi, and Martin Savage have claimed that such a spacetime grid could be revealed based on observing the highest energy cosmic rays – which I suspect is likely to convince zero cosmic ray physicists, regardless of the outcome of such a study.

There is, moreover, a flaw in relying on any test involving observations of the physical universe at large. Such tests would be meaningless if the simulation you are living in does not involve the whole universe, but simply that tiny portion of it that *you* perceive – see Figure 6.11. This possibility, originated by philosopher Gilbert Harman, has been called a "brain in a vat," and it is the theme of many science fiction stories. According to

FIGURE 6.11
Brain in a vat believes it is walking, image created by Alexander Solberg Wivel.

such stories, a computer is simulating reality, and the disembodied brain to which it is connected continues to have perfectly normal conscious experiences, making it think it observes a real world. Obviously, the complexity of a simulation in this case would be enormously simpler compared to one made for the whole universe. Perhaps, the best hope of discovering we are just a "brain in the vat" might be to look for "magical" events that defy explanation.

My Belief That I Am Living in a Simulation

I normally do not have much use for "New Age" mystical thinking, but there were three remarkable events occurring during my tachyon research, that strongly reinforced my sense of living in a simulation. The first event, discussed at length in Chapter 1, was when I received what appeared to be an automated suggested revision in the title of a paper I had just submitted to a journal. This suggested revision could not have come from the journal's online submission system, according to the journal editor, and I have no clue from where it originated. Whatever its source, this "message" encouraged me that I was on the right track with my research, and it even made me wonder if I could possibly have received a tachyon message from the future.

The second event was the almost magical appearance of an 8 MeV neutrino line in the Kamiokande SN 1987A data. Even though the existence of this line is not yet firmly established, this "unicorn" appeared in the data quite suddenly. The instant I realized that I could use data taken by the detector at a different time as the background spectrum for the data taken on the day of SN 1987A, I saw that all the criteria for a true neutrino line were met. What I had formerly assumed was a spectrum consisting of only background events suddenly became a neutrino line sitting atop the background. The data did not change, but my perception of it suddenly changed drastically, much as occurs with "ambiguous images" – see Figure 6.12.

The third mysterious event has not been discussed yet. The event occurred when I was seeking confirmation of my Z′-mediated reaction model to explain an 8 MeV neutrino line from SN 1987A. As discussed in Chapter 3, this model required the existence of dark matter X particles of 8 MeV mass, which might also be present near the galactic center. By accident, I had come across a 2011 paper by astrophysicist Nicolas Prantzos and collaborators on the spectrum of gamma rays from the galactic center. This paper, using gamma-ray data from four instruments, also showed computed spectra for dark matter annihilations using five X-particle masses: 0, 5, 10, 50, and 100 MeV. Commenting on their graph, the authors noted that any X mass greater than "a few MeV" could be ruled out by the data. If that statement were true,

FIGURE 6.12
An example of an ambiguous image. Do you see a tree or two faces?

Source: Image from Shutterstock

it would, of course, rule out my model involving X particles of mass 8 MeV. Something about their graph, however, made me want to pursue the situation further. While trying to understand better how different values of the X-particle mass affected the shape of the gamma-ray spectrum, I came across other gamma ray data recorded by an instrument known as OSSE that had not been included in the Prantzos graph.

Now here is the magical part. When I combined the OSSE data points with those displayed by Prantzos into a single graph the result was astounding. The Prantzos data and the OSSE data each separately were quite consistent with an X-particle mass of zero, that is, no sign of dark matter. However, the combination of the two datasets remarkably showed clear evidence for a specific non-zero X mass, $m_X = 10$ MeV. In fact, the 10 MeV curve that Prantzos displayed for that mass value yielded near-perfect agreement with the combined data! By generating a few other curves, I found that, the combined data gave good fits to any X-particle mass in the approximate range 8 to 15 MeV, an interval that barely included my predicted ~ 8 MeV mass value. In summary, what had started out as an apparent refutation of the hypothesis of a dark matter origin of galactic center gamma ray data remarkably turned into an important validation of it.

As I noted earlier, the preceding three strange occurrences during my research have increased my sense of living in a simulation. Let me pursue this speculation one ridiculous step further. Suppose I should be lucky enough to see my 3 + 3 model verified either by the KATRIN experiment, or even more remarkably, by a new supernova in our galaxy. Were such an event come to pass during my lifetime, it would exponentially increase my belief that I am living in a simulation, because it would likely mean I am not the poor delusional schlub who just thought he discovered tachyons!

Some Concluding Thoughts

The main advice offered in this chapter has been that to make an important scientific discovery, you should consider the method of data prospecting as an alternative to the more traditional methods of theory or experiment. In fact, I believe that this method is superior to theory alone if the theories are completely divorced from any experimental predictions or even any concern about the need for such predictions, as has been the case with string theory. Other suggestions are subsidiary to this one, and they are summarized below.

Finding a worthwhile topic for data prospecting:

- Keep an open mind about the limits of the possible.
- Unfashionable or controversial topics have much less competition.
- The best topics are those most people dismiss for questionable reasons.
- Remember the "Alvarez rule" about the worthiness of a research topic.
- Keep an eye out for datasets that show promising anomalies.

Developing and testing your model:

- Do not expect every model derived from data to be fruitful.
- Look for alternative assumptions in interpreting experimental data.
- Aim for models that have very specific and multiple testable predictions.
- Strive for consilience in seeking many kinds of data to support the model.
- Implausible models become more plausible after each test passed.
- Watch out for negative evidence that might doom your model.

- Abandon or modify a model unless you can rebut the negative evidence.
- Every empirical test that does not kill your model only makes it stronger.
- Avoid the invention of *ad hoc* explanations to save a model.

Other advice:

- A lengthy break may provide a new perspective on your previous work.
- Realize that mistakes, even truly atrocious ones, are inevitable.
- Learn from your mistakes and negative reviews of your papers.
- Even ill-informed negative reviews may contain useful feedback.
- Learn to identify fake journals and conferences and avoid them.
- Avoid self-publishing and sending people unsolicited manuscripts.
- Seek out collaborators whose expertise is complementary to yours.
- If you lack a collaborator, find someone to give critical feedback.
- Do not always believe the "experts" if your intuition tells you otherwise.
- Reach out to experimental groups that can test your model.

You should be aware that some of the advice summarized here is contradictory. Thus, when you avoid the invention of *ad hoc* explanations to save a model, you need to be careful that you are not dismissing an idea for which you might find evidence if you are more open minded about the limits of the possible. Of course, you must be willing to abandon a pet theory if irrefutable negative evidence arises. However, if you become convinced that you are on the right track, never give up in your pursuit of the evidence for your idea, *and* negative evidence against it.

Summary

This chapter summarizes my advice to others who seek to make an important discovery in physics or other sciences. Most importantly, it discusses the method of data prospecting as an alternative to theory or experiment. Other suggestions are subsidiary to this one, and they include how to find a worthwhile topic for doing data prospecting, and what makes for a particularly attractive topic. One important possible topic discussed at length is the SETI. The multiple avenues for such a SETI search include looking for: (a) non-natural signals from space (b) ways to communicate with other intelligent species on Earth, (c) signs of an object from outside the solar system having an artificial origin, (d) signs of Dyson Spheres around stars, and (e) signs of ET civilizations that have destroyed themselves. The chapter also includes much advice (some of it contradictory!) on how to validate an unconventional model arising from an analysis of data. Thus, making up *ad hoc* explanations to save a model which fails to fit new data should be avoided, but in such cases, you need to be careful that you are not dismissing an idea for which you might find evidence if you are more open minded about the limits of the possible.

I therefore offer encouragement to all readers, who might wish to become data prospectors. Go forth boldly where no one has gone before and develop and test your models that defy conventional wisdom. Do not be discouraged if others ridicule what you find, especially if you keep finding new evidence, and no fatal negative evidence emerges, but if it does, move on to something more fruitful. As it has been wisely observed "All truth passes through three stages. First, it is ridiculed. Second, it is violently opposed. Third, it is accepted as being self-evident."[11] So, if your model should get published, hopefully not in a fake journal, and then become widely ridiculed, I offer my congratulations on your making it to second base. With some luck you may then find your model violently opposed, and from there, who knows?

References

1. The adage appears to be originally due to Spanish Essayist, novelist, poet, playwright, and philosopher Miguel de Unamuno, but variations of it have been attributed to many others, including Albert Einstein.
2. Doyle, Arthur Conan. *The Sign of the Four,* New York, NY, Integrated Media, p. 111, 1890.
3. Clarke, Arthur Conan. *Hazards of Prophecy: The Failure of Imagination,* first published in *Profiles of the Future,* New York, NY, Bantam Science and Mathematics, 1962.
4. Kragh, Helge. *Max Planck: The Reluctant Revolutionary,* Physics World, December 1, 2000.
5. Wescott, Roger W., "Anomalistics: The Outline of an Emerging Field of Investigation," edited by Maruyama, M., and A. Harkins, *Cultures Beyond the Earth,* New York, NY, Vintage Books, pp. 22–25, 1975.
6. Klass, Philip J., The Skeptics UFO Newsletter, vol. 32, March 1, 1995.
7. Stapledon, O., Star Maker, London, Methuen Publishing, 1937.
8. Jones, Alexander. *Like Opening a Pyramid and Finding an Atomic Bomb,* Proceedings of the American Philosophical Society, 162, no. 3, 259–294, 2018.
9. Shklovsky, Iosif, and Carl Sagan. *Intelligent Life in the Universe,* Boca Raton, Palm Beach, FL, Emerson Adams Press, 1966.
10. Bostrom, Nick. *Are You Living in a Simulation?* Philosophical Quarterly, 53, no. 211, 243–255, 2003.
11. This quote is usually attributed to Arthur Schopenhauer (1788–1860), but its true origin is unclear.

Index

Note: Locators in *italics* represent figures and **bold** indicate tables in the text.

A

Absolute future, 37
Absolute past, 37
Acceleron, 103
Aether, 122
Alcubierre, Miguel, 111
Alcubierre Warp Drive, 111–112, *112*
Alvarez rule, 188
ANITA Collaboration, 117
Anomalistics, 173
Antares, 69
Antigravity, 115–116
Antikythera mechanism, 180, *180*
Antimatter, 139
Antineutrinos, 19, *21*, 41–42, 137
Antiparticles, 19, 101
Antiquarks, 101
Antitelephone, 52
Arago, Francois, 161
Arago/Poisson spot experiment, *161*
Atomic bomb, 3
Atomic nucleus, 11
Autiero, Dario, 148
Axino, 103
Axion, 103

B

Background spectrum, 89–91
  reliability of, 91–93
Backward time-traveling tachyons, 37–39
Bayes Rule, 89
Behroozi, Peter, 107
Bell's theorem, 121–122
Bem, Daryl, 26
Benford, Gregory, 31, 52
Berners-Lee, Tim, 163
Beta decay, 18, 41
Beta particle, 18
Betelgeuse, 69
Big Bounce theory, 118–119
Bilaniuk, O.M.P., 4, 28
Bilepton, 103
Black dwarf, 71
Black holes, 115
  electron, 103
Black Swan: The Impact of the Highly Improbable, The (Taleb), 17
Black swan phenomenon, 17
Block universe, 35
Bojowald, Martin, 118
Borissova, D., 117
Bosons, 101
Bostrom's assumptions, 183
Box, George, 21
Bradyons, 6
Brief Answers to the Big Questions (Hawking), 57
Butler, A. H. Reginald, 9

C

"Canals" on Mars, 15
Carroll, Sean, 99
CASA-MIA, 48
CERN accelerator, 9
CERN 2m hydrogen bubble chamber, *134*
Chameleon, 103
Chargino, 103
Cherenkov, Pavel, 110–111
Cherenkov radiation, 110
Chodos, Alan, 41, 126, 152
Chronology protection conjecture, 57
Clarke, Arthur C., 28
Clay, Roger, 23
Cohen, Andrew, 127
Coker, Rory, 26
Colladay, Don, 125
Communicative intelligent civilizations, 178

Constellation Aquarius, 124
Coronavirus pandemic, 152
Cosmic inflation, 114–115
Cosmic microwave background (CMB) radiation, 124, 131–132
Cosmic rays, 42–43, *43*, 51, 72
- flux *versus* energy, 44
- GZK cutoff, 50
- particles, 23
- physics, 50–51
- protons, 47
- sources of, 45–46, 49
- spectrum, 78

Cosmology, alternative to, 118–119
Cosmon, 103
Cowan, Clyde, 19, 95
Crab Nebula, 67, *68*
Crouch, Phillip, 23
Cryonics, 109
Crypton, 103
Curie, Marie Skłodowska, 3
Curvaton, 103
Cygnus X-3, 46–48, 50, 63

D

D'Alembert, Jean, 33
Dark energy, 104, 115–116
Dark matter, 80–81, 104
- and accelerated expansion, 116–117
- in stellar core and 8 MeV neutrinos, 84–86

Dark matter X-particles, 84–85, *85*
Dark photon, 103
Data prospecting, 2, 165–166
Da Vinci, Leonardo, 26
Delisle, Joseph-Nicola, 162
Denis, Siméon, 159
Dent, James, 26
Deshpande, V.K., 4, 28
Dilaton, 103
Diphoton, 103
Dorado constellation, 73
Doyle, Arthur Conan, 164
Drake, Frank, 178
Drexlin, Guido, 147, 149
Dual graviton, 103
Dual photon, 103
Dyson, Freeman, 177–178
Dyson spheres, *177*, 177–178, 189

E

Earthlings, 15
Edmunds, Mike, 180
Effective mass, 78
Eight MeV neutrino line, 93–94, 96, 186
Einstein, Albert, 3, *4*, 28, 31–33, 100, 105, 119, 165
- ether, 125–126
- geometric formulation, 108
- relativity, 37, 109
- Ring, *59*
- Special Theory of Relativity, 1–3, 31–32, 100
- theory of gravity, 56

Electromagnetic forces, 100
Electromagnetism, 100
Electrons, 41, 131, 137
- energy, 137
- neutrino, 139

Elementary particles, standard model of, *102*
Elkins, Geoff, 136
Elsewhere, 37
$E = mc^2$, 3, 18, 32, 70, 85
Energy, 41
Ereditato, Antonio, 148
Erlykin, Anatoly, 45
Ether, 122
Exploding stars, threats, 68–69
Extraterrestrial civilization, 176

F

"Face" on Mars surface, 15, *16*
Faster than light (FTL) particles, 1, 5, 31, 74, 103
- Einstein's prohibition of, 2–4
- neutrinos, 9, 128
- observers and warp-drive spaceships, 111–114
- speed, 1–3
- tachyons, 5, 8–9, 148

"Fast-living" supergiants, 69
Feinberg, Gerald, 109

Feng, Jonathan, 104
Fermi, Enrico, 13–14, 18–19, 163
Fermi National Lab, *164*
Feynman, Richard, 99, *100*, 101
FitzGerald, George, 110
Flandern, Tom Van, 16
Forbidden beta decay of proton, 41
Force carriers, 101
Ford, Ken, 81
Forgan, Duncan, 181
Frank, Illya Mikhailovich, 111
Freeth, Tony, 180
FTL, *see* faster than light (FTL) particles

G

Galactic supernova, 79
Galileo Project, 176
Gamma-ray bursts (GRBs), 88, *88*
Gates, Evalyn, 41
Gaussian, 93
Gedankenexperiments, 31
General relativity, 56
Ghez, Andrea Mia, 86, *87*
Glashow, Sheldon, 127
Gluino, 103
G–2 muon anomaly, 104–105
God, 94–95
God Particle, 14
Goldstino, 103
Gonzalez-Mestres, Luis, 51
Graviphoton, 103
Graviscalar, 103
Gravitational forces, 100
Gravitational lensing, *59*
Gravitino, 103
Graviton, 6, 103
Gravitophobic neutrinos, 131–132
  beta decay spectrum, 137–139
  beta spectrum shape for tachyon, 140–141
  controversial results, 147–148
  data, 146–147
  electron energy in spectrometer, 145–146
  electron neutrino mass, 137–139
  imaginary mass, scale for measuring, 135–136
  KATRIN and tachyons, 153–154
  KATRIN experiment, 143–145
  KATRIN initial data with 3 + 3 model, 150–152
  muon neutrino, weighing, 132–135
  neutrino mass experiments, 141
  spectrum, 139–140
Gray, Kathryn, 71
Gray, Paul, 71
Great Depression, 182
Grossmann, Marcel, 32
Guagino, 103
Guth, Alan, 114
GZK cutoff, 50

H

Harman, Gilbert, 185
Hawking, Stephen, 57, *58*, 118
Heaviside, Oliver, 109–110
Heisenberg Uncertainty Principle, 58
Helium, 90
Helium fusion, 70
Helium-3 nucleus, 137
Henderson, Bobby, 34
Herbert, Don, 10
Hess, Victor, 50
Higgs boson, 14
Higgs field, 14, 102
Higgsino, 103
Higgs ("molasses") field, 124
Higgs particle (H), 101, 103
Hirata, Keiko, 89
Holium, 103
Hoyle, Fred, 114
Hubble, Edwin, 114
Huygens, Christiaan, 162
Hydrogen, 70
Hydrogen fusion, 70
Hypothetical particles, 103
Hypothetical tachyons, 6

I

IceCube, 125
IceCube Neutrino Observatory, 125, *126*
Infinite energy, 7
Inflationary expansion, 114

Inflation era, 58
Integral spectrum, 146
Intelligent Life in the Universe (Shklovsky and Sagan), 181
Interstellar dust, 69

J

James, Jack O'Malley, 181
Jentschura, Ulrich, 126
John Wheeler, 58

K

Kamiokande data, **160**
Kamiokande-II (K-II), 72, 74
  collaboration, 89, 91
  detector, 83–84, 96
  neutrinos, 90
Kamiokande SN 1987A data, 186
Kaon, 101
KArlsruhe TRItium Neutrino (KATRIN) experiment, 2, 23, 95, 141, 143–145, 149, 151, 172
Knee of the spectrum, 43, 46
Kostelecký, Alan, 41, 125
Krasznahorkay, Attilla, 84, 104
Kulp, Daniel T., 27–28

L

Lane, Dave, 71
Large Hadron Collider (LHC), 106
Law of energy conservation, 19
Lederman, Leon, 12, 104
Leptoquark, 103
Liao, Jiajun, 125
Light cone, 36
Loop quantum cosmology, 108
Loop quantum gravity (LQG), 107–108
Lorentz, Hendrik, 37, 124
Lorentz FitzGerald contraction, 110
Lorentz force, 110
Lorentz transformation, 37–38, *38*
Lorentz violation, 125
Loyalty Oath, 13
Lucas, John, 35
Luxons, 6

M

Mack, John, 175
Magnetic moment, 104
Magnetic monopole, 103
Magnetic photon, 103
Majoran, 103
Maldacena, Juan, 57, 121
Mallet, L. I., 111
Man Ho Chan, 82
Maraldi, Giacomo Filippo, 162
Marfatia, Danny, 125
Martian canals, 15
Martians, 15
  civilization, 15
Mass, 5
Matrix, The, 182
Maxwell, James Clerk, 100
Meta-particles, 110
Meta relativity, 4–7, 62
Michelson, Albert, 123
Michelson–Morley experiment, *123*, 124
Micro black hole, 103
Milekhin, Alexey, 57
Milky Way galaxy, 46, 67, 80
Milner, Yuri, 176
Minkowski, Hermann, 32
Mirror universes, 117–118
Molasses, 102
Mont Blanc detector, 74, 130
Mont Blanc neutrino burst, 82–84, 89, 94, 96, 159–160, 171
Morley, Edward, 123
Morris, Michael, 62
M-theory, 106–107, 129
Multiverse, 107
Muons, 12, 104, *134*, 135
  kinetic energy, 134
  neutrino, 132–136, 158
  neutrino experiment *see* two-neutrino experiment

N

Nahin, Paul, 110
Nakahata, Masayuki, 89
Neutralino, 103
Neutrinos, 1–2, 14, 18–21, *21*, 45, 71, 101, 103–104

bursts, 73
detection, 72–73
detector, 83–84
energy, 73, 93
FTL tachyons, 8–9
history, 20
hypothetical neutrino, 19
masses, *134*, 135, 141
mass hierarchy, 22
mass hierarchy, higher energy, 24
8 MeV neutrino line, 93–94
Mont Blanc neutrino burst, 82–83
negative energy, 41
neutrino burst, 73–75
oscillations, 19–20, 75
shock wave of light incident on photo-sensors, *73*
SN 1987A neutrino burst, 73–75
speed, 9
"sterile" (inert), 20–21
tachyon, 47
weighing, 158
Neutron, 45
Newton, Isaac, 100, 162
Newton's particle theory of light, 159
Niven, Larry, 55
Niven's Law of time travel, 55
Non-Communist Oath, 12
Novikov, Igor Dmitriyevich, 54
Nuclear fusion, 69
Nye, Bill, 10

O

Ockham's Razor, 24
OPERA
beam, 128
Collaboration, 148
group, 8–9
Orion, 69
Oumuamua, 173–174, *174*, 175–176
Ozone layer, 68

P

Pakvasa, Sandip, 26
Panpsychism, 35
"Parent" and "daughter" nuclei, 18
Parkes Observatory, 177
Parno, Diana, 152
Particle Data Group, 6
Particle physics, standard model, 100–102
Pas, Heinrich, 26
Pastafarianism, 34
Pauli, Wolfgang, 18–20, 95, 103
Perlmutter, Saul, 117
Photino, 103
Photon, 37, 45, 102
Photo-sensors, 73
Pion, 101
Pion decay, 132–133, *134*, 136
Planck, Max, 165
Planck length, 58
Planck particle, 103
Plasma globe, *11*
Pleckton, 103
Podolsky, Boris, 119
Poisson, Siméon Denis, 159–161
Polly, Chris, 105
Pomeron, 103
Post hoc analysis, 16–17
Potting, Robertus, 41
Prantzos data, 187
Precognition, 24–25
Predatory journals, 168–169
Preon, 103
Pressuron, 103
Prometheus Project, The (Feinberg), 109
Proton, 41
Proton beta decay, 41
Proton-neutron decay chain, 46–47, *47*
Pythagorean theorem, 7

Q

Q-ball, 103
Quantum entanglement, 119–121, 164
Quantum field theory, 109
Quantum foam, 58
Quantum mechanics, 20
Quantum of Higgs field, 102
Quantum theory of gravity, 107–108
Quantum theory of matter, 101
Quarks, 101–103
Q-value of tritium, 137

R

Rabounski, Larissa, 117
Radioactivity, 72
Radioisotope, 137
Radzikowski, Marek, 126
Realism, 120
Recami, Erasmo, 111
Red galaxy LRG 3–757, *59*
Reines, Frederick, 19, 95
Relativistic mass, 5
Riess, Adam, 117
Rosen, Nathan, 119
Rovelli, Carlo, 107–108
Rubin, Vera, 81, *81*
Rujula, Alvaro de, 50
Rutherford, Ernest, 11–12

S

Sagan, Carl, 67, 181
Samuel, Stuart, 125
Sanduleak-69 202, 67, 73
Saxion, 103
Schiaparelli, Giovanni, 15
Schmidt, Brian, 117
Schrodinger, Erwin, 35
Schrodinger's cat, 120
Schrodinger's mouse, 120
Schwartz, Charles, 127
Schwartz, Melvin, 12
Science of Interstellar, The (Thorne), 61
Scorpius, 69
Search for extraterrestrial intelligence (SETI), 173
Seesaw analogy, 78–79
Selectrons, 106
SETI data, 176, 189
Sfermion, 103
Sgoldstino, 103
Shklovsky, Iosif, 181
"5-sigma" rule, 135
Silurian Hypothesis, 181
Singularity, 183
Skyrmion, 103
Smolin, Lee, 107–108
SN 1987A, 67, 71, 75, 82, 91, 93, 162
  3 + 3 model of neutrino masses, 75–77
  neutrino burst, 73–75
  neutrino masses, 78
  neutrinos, 88–89, **160**
  spectrum, 83
  three unicorns associated with, 96–97
SN 1987 B, 67
Snow, C. P., 32
Solar eclipse, *170*
Solar supernova, 68
Solar system, 15
Sommerfeld, Arnold, 4, 109–110
Spacetime, 33
  absolute future, 37
  absolute past, 37
  elsewhere, 37
Sparticles, 106
Special Theory of Relativity, 31–32
Spectrum endpoint, 137
"Spin zero" tachyons, 113
Squarks, 106
Stars
  birth, 69
  burnt-out cinder, 71
  core of, 70, *70*
  death, 70–71
  life, 70–71
  Sanduleak-69 202, 73
  supernovae, 69
Steinberger, Jack, 12–14
Steinhardt, Paul, 118
Stellar explosions, 67
Sterile neutrino, 20, 103
Stevens, Adam, 181
String theory, 105–106, *108*
Strong forces, 100
Subatomic particles, 1, 6, 40
  reactions, 133
Sudarshan, Ennackal Chandy George, 4, *5*, 28, 110, 167
Super-atom, 103
Supercomputer, *164*
Superdeterminism (SD), 121–122, *122*
Supergiants, 69
Super-Kamiokande detector, 72, *72*, 110
Superluminal neutrinos, 9, 14, 148
Superluminal speed, 1, 5, 8, 128

Supernovae, 67–69
SuperNova Early Warning System (SNEWS), 71
*Super*string theory, 106
Supersymmetry, 106
Susskind, Leonard, 121
SUSY, 106–108, 129

T

Tachyon, 103, **160**
Tachyonic neutrinos, 77–78, 126–127, 140, 153–154
Tachyons, 1, 5–6, 8–9, 14–15, 28, 115–116
  backward time-traveling tachyons, 38–39
  chasing, 39–40
  consciousness, 40
  direction changes, 39
  direction of motion, reversal, 40–41
  fact and fiction, 31
  gun, 40
  messaging, 52–55
  possibilities, 23
  and wormholes, 62
Taleb, Nassim Nicholas, 17
Tamm, Igor Yevgenyevich, 111
Tardyons, *see* bradyons
Tau neutrino, 103
Teleportation, 164
Theory of everything (TOE), 105
Thermonuclear explosion, 71
Thomson, William, 81
Thorne, Kip, 61–62
Three neutrino masses, models, 21–23
3 + 3 model of neutrino masses, 27, *75*, 75–77, 94, 132, 140–142, 149, 159, **160**, 188
  confirmation of, 82
  dark matter, 80–81
  Mont Blanc neutrino burst, 82–83
  neutrino detector, 83–84
  seesaw analogy, 78–79
  SN 1987A, 75–77
  tachyonic third mass, 77–78
  validation for, 79
Three unicorns associated with SN 1987A, 96–97, *97*
Threshold energy, 78
Time, perception of, 33
Time as fourth dimension, 33
Timescape (Benford), 152
Time travelers, 57–59
Tripartite classification scheme, 6
Tritium, 137, 139
Tritium beta decay, 137, *137*
Troitsk, 149
  anomaly, 142
  Collaboration, 142
Tunka Cosmic Ray Collaboration, 49
Two-neutrino experiment, 12–14, 132

U

Ufology, 173
Ultra-heavy electrons, 12
Universal speed limit (c), 7

V

Van Helmont, John Baptist, 147
Vavilov, Sergei, 111
Verne, Jules, 26
Vibrating strings and membranes, 105–107
Von Neumann, John, 35, 154–155

W

Wave theory, 161
Wave theory of light, 2
W′ boson, 103
Weak forces, 20, 100
Weiler, Thomas, 26–27
Wells, H. G., 26, 33
Wescott, Roger W., 173
Wigner, Eugene, 35
William of Ockham, 24
Witten, Edward, 105
Woit, Peter, 107
Wolfendale, Arnold, 45
Worldline, 33–37
  and light cone, *36*
  spacelike, 37
  timelike, 37

Wormholes, 59–60, *60*
  in solar system, 60–62
  and tachyons, 62
  time machines, 56–57

X

X-particle ignition temperature, 85

Z

Zahn, Corvin, 59
Z′ boson, 103
Zero-mass neutrino, 139–140
Z′-mediated reaction model, 84, 86–87, 89, 96
  challenges, 87–89
Zotov, Mikhail, 49
Zwicky, Fritz, 81